Energy Systems in Electrical Engineering

Series editor

Muhammad H. Rashid, Pensacola, USA

More information about this series at http://www.springer.com/series/13509

Rohit Dhiman · Rajeevan Chandel

Compact Models and Performance Investigations for Subthreshold Interconnects

Springer

Rohit Dhiman
Rajeevan Chandel
Electronics and Communication Engineering
National Institute of Technology
Hamirpur, Himachal Pradesh
India

ISSN 2199-8582 ISSN 2199-8590 (electronic)
ISBN 978-81-322-2997-1 ISBN 978-81-322-2132-6 (eBook)
DOI 10.1007/978-81-322-2132-6

Springer New Delhi Heidelberg New York Dordrecht London

Softcover reprint of the hardcover 1st edition 2015

Printed on acid-free paper

Springer is part of Springer Science+Business Media (www.springer.com)

To my parents

Rohit Dhiman

To my family

Rajeevan Chandel

Preface

Modern very-large-scale integration (VLSI) chips contain millions of transistors. It is envisaged that VLSI chips will contain more than three billion transistors in the coming decade. VLSI chips find wide applications in all modern electronic circuits and systems. In highly-scaled VLSI technology nodes, more and more functionalities are being housed on a chip. Various functional blocks of the chip are connected to each other with interconnects. Interconnects distribute clock, supply voltage, and signals in a VLSI chip. Chip sizes are also decreasing due to technological advances. High chip complexity requires dense interconnections to communicate information between devices and circuit blocks. As a result, long interconnects have become common on-chip features. However, long interconnects cause various deleterious effects, viz., high propagation delay, degradation of signal waveforms, excess power dissipation, and crosstalk. Consequently, the increased number of interconnects in present-day VLSI chips has made the interconnection design problem complex and challenging.

In recent years, power dissipation is given comparable weightage to the area and speed considerations. The primary driving factor is increasing prominence and fast growth of battery operated applications such as micro-sensor networks, pacemakers, hearing aids, and many other portable devices, which require stringent energy constraints for longer battery lifetime. Subthreshold operation of devices presents an opportunity for energy-constrained applications with its ultra-low-power consumption. Subsequently, the benefits from ultra-low-power operation have carved out a significant niche for subthreshold circuits. Though the subthreshold operation shows huge potential toward satisfying the ultra-low-power requirements of portable systems, it holds design issues both for interconnects and circuit design. These issues lead to significant increase in the design complexity of integrated circuits. There are not many works that address these challenges for subthreshold circuit design in an integrated and comprehensive manner. This book provides a detailed analysis of concerns related to subthreshold interconnect performance from the perspective of analytical approach and design techniques. It also presents a qualitative summary of the work reported in the literature by various researchers in the design of digital subthreshold circuits. Particular emphasis is laid on the performance analysis of coupling noise and variability issues in subthreshold domain to develop efficient compact models. The different tasks accomplished in this book are mentioned below.

A new parameter called subthreshold drain conductance is defined. Two new subregions of MOS operation are identified. Compact analytical expressions governing output voltage, propagation delays, and coupling noise are developed. The impact of coupling on aggressor delay is analyzed. The proposed analytical approach gives physical insight into the parameters affecting the transient behavior. This is essential for avoidance of dynamic crosstalk and circuit malfunctioning. Remedial design techniques are suggested to mitigate the effect of coupling noise caused by the interconnect coupling capacitance. The effects of wire width, spacing between the wires and wire length are thoroughly investigated. In addition, the effect of parameters like driver strength on peak coupling noise has also been analyzed. Process, voltage, and temperature variations are prominent factors affecting subthreshold design and have also been investigated. Analytical expressions characterizing variability based on the parametric analysis are developed. The process variability analysis has been carried out using parametric analysis, process corner analysis, and Monte Carlo technique. The impact of temperature on subthreshold interconnect performance is also investigated.

To summarize, this book will serve as a platform for researchers and graduate students with deeper insights into subthreshold interconnect models in particular and designing complex logic gates in general. This book will best fit as a textbook and/or a reference book for students who are initiated in the area of research and advanced courses in nanotechnology, ultra-low-power interconnect design, and modeling.

Rohit Dhiman
Rajeevan Chandel

Contents

About the Authors

Dr. Rohit Dhiman received his B.Tech. degree in Electronics and Communication Engineering from H.P.U. Shimla, India in 2007. He completed his M.Tech. in VLSI Design Automation and Techniques from National Institute of Technology (NIT) Hamirpur, India in 2009. He was awarded the Ph.D. degree from NIT Hamirpur in 2014. He collaborated with Nanoelectronics Research Group at Indian Institute of Technology (IIT) Ropar, India for Post-Doctoral Research. Presently Dr. Rohit Dhiman is working as Assistant Professor at NIT Hamirpur. He has over 20 research papers in international journals of repute and conferences to his credit. His research interests include device and circuit modeling for low power VLSI design.

Dr. Rajeevan Chandel received her B.E. degree in Electronics and Communication Engineering from Thapar Institute of Engineering and Technology, Patiala, India in 1990. She is a double gold medalist from Himachal Pradesh University, Shimla in Pre-University and Pre-Engineering. She completed her M.Tech. in Integrated Electronics and Circuits from IIT Delhi in 1997. She was awarded the Ph.D. degree from IIT Roorkee, India in 2005. Dr. Chandel joined Department of Electronics and Communication Engineering, NIT, Hamirpur, HP as Lecturer in 1990, where presently she is working as **Professor**. She has over 40 research papers in international journals of repute and over 90 in conferences. Her research interests include electronics circuit modeling and low power VLSI design. She is a life member of IETE (I) and ISTE (I) and a member of VSI.

Chapter 1
Introduction

Keywords Complementary metal oxide semiconductor (CMOS) · Deep submicron · Interconnects · Subthreshold · Very-large-scale integration (VLSI)

New and extensive innovations in solid-state very-large-scale integration (VLSI) technology have led to the revolution in high-speed networks, communication devices, and a host of many other electronic equipments in the present era. VLSI chips find wide applications in all modern electronic circuits and systems. Further, VLSI technology has reduced the voluminous electronic parts which were used to manufacture early day's electronic equipment and computing machines. Technological advances in VLSI have led to downsizing the device size and weight and increased the reliability, thus facilitating miniaturization of the size of electronic devices and circuits. VLSI functionality along with data computation has enhanced the capabilities of the electronic gadgets. Miniaturization has directly or indirectly been the core cause of the tremendous applications of integrated circuits (ICs) and their omnipresence in all electronic systems. The decrease in minimum feature size and increase in number of gates in a chip area have been due to technology scaling. Furthermore, improvements witnessed in various present-day electronic gadgets have been mostly attributed to advances in VLSI technology. As a result, both die size and device densities of the circuits have increased.

1.1 Preliminary Background

In 1958, Jack Kilby at Texas Instruments conceived the design of the first IC on a single substrate [1]. Presently, complementary metal oxide semiconductor (CMOS) technology has advanced the ICs way forward. It has led to the astounding evolution toward higher integration densities. The level of integration has evolved from small-scale integration (SSI) to VLSI with nearly 10^6 devices/chip, ultra-large-scale integration (ULSI) with more than 10^7 devices/chip, and giga-scale integration (GSI) having more than 10^9 devices per chip [2]. With current GSI level, more functional blocks are being included on a single chip. Various functional blocks of the chip are connected to each other with interconnects. The global or long

R. Dhiman and R. Chandel, *Compact Models and Performance Investigations for Subthreshold Interconnects*, Energy Systems in Electrical Engineering, DOI 10.1007/978-81-322-2132-6_1

interconnects distribute clock and signals and provide power and ground to billions of active devices on a chip. Moreover, VLSI chip has a single layer of transistors and may have more than seven to nine layers of interconnections [3]. As per the international technology, roadmap for semiconductors (ITRS), the future nanoscale circuits will house more than a billion transistors [4]. With such high chip complexity, i.e., increase in the number of active devices per chip, long interconnects have become common on-chip features in the recent high-density circuits. Thus, size of the chip and functionality on the chip have increased manifold, thereby increasing the length of interconnects within the chip circuitry.

With appreciable lengths of interconnects in VLSI chips, the associated interconnect impedance parameters viz. resistance, capacitance, and inductance have increased significantly. These parameters have a direct impact on the system performance and cause quadratic increase in signal propagation delay. Interconnects also cause excessive power dissipation due to the associated parasitic impedance elements [5, 6]. Up to 30 % of the dynamic power is consumed by the global interconnect network [7]. Global interconnects in nanometer technologies which connect billions of active devices on a chip have therefore become major showstoppers. Consequently, the effect of global interconnect impedance parameters on delay and power dissipation is of great concern for the VLSI circuit designers.

In addition to delay and power, global interconnects pose another design challenge due to decreasing feature size and increasing length of on-chip interconnects for highly scaled technology nodes. In deep submicron (DSM) circuit design, typically less than 0.18 μm, the spacing between interconnects is reduced. This in turn increases the coupling capacitance and becomes comparable to the line-to-ground interconnect capacitance [8]. Interconnects thus exhibit coupling which is the major source of coupling noise and is commonly referred as crosstalk. Coupling noise causes deleterious effects such as high propagation delays and power dissipation and induces overshoots and undershoots. Crosstalk-induced overshoot and undershoot can generate and propagate false switching in the circuit. The effect of this unwanted interference depends on the value of the coupling capacitance, signal transitions, and the adjacent interconnect length. Consequently, interconnect coupling noise has become one of the primary threats to the continued growth in integration density [9]. Therefore, it is very important to predict the interconnect noise at the system level as it can create a logic error at the noise site [10].

In the early 1990s, the primary concerns of the VLSI designers have been area, performance, cost, and reliability. Power consideration had been mostly of secondary importance. However, with limited battery or to enhance the battery life, power has drawn a comparable attention of the designers along with the area and speed considerations. The primary driving factor has been the remarkable success and fast growth of portable battery-operated ultra-low-power applications such as wireless sensor networks, self-powered radio frequency identification, wearable battery-powered systems, and implantable circuits for biomedical applications. Portable devices are bound by ultra-low-power budget, which places a pressing demand of ultra-low-power operation for a longer battery lifetime. Reduced power consumption makes the circuits lighter, reduces or eliminates cooling subsystems, reduces the weight, and extends the life of the energy source. The demand for ultra-low-power circuits in CMOS IC design has thus become an important design challenge for next-generation DSM technologies.

It has been observed that the subthreshold operation of the device exhibits a great potential toward satisfying ultra-low-power demand of portable systems [11–13]. Subthreshold operation uses supply voltage which is below the transistor threshold voltage. Driving CMOS circuits with the subthreshold current provides orders of magnitude power reduction over strong inversion circuit operation. As a result, the subthreshold logic has emerged as an important approach to design energy-efficient systems [14]. The subthreshold current is much lower in magnitude than the saturation drain currents in the strong inversion regime. The weak driving current inherently limits the performance but minimum energy is achieved, resulting in a longer battery lifetime. Consequently, ultra-low-power operation of MOS devices has drawn the attention of researchers in recent years [15–20].

Furthermore, with each scaled semiconductor technology generation, the impact of process, voltage, and temperature (PVT) variations on the subthreshold system performance has significantly increased [21]. This is because of the exponential dependency of subthreshold current on the device threshold voltage and temperature. Small changes in the threshold voltage and temperature translate into exponential variations in the bias current, thereby affecting the device delay and power dissipation. Thus, it is important to investigate the influence of PVT variations under subthreshold regime on the interconnect performance along with devices while designing robust ultra-low-power systems.

In the research papers available in the literature regarding subthreshold circuits, researchers have investigated the performance of these circuits in terms of speed and robustness either by device- or circuit-level optimization [22]. Gate voltage boosting technique to improve subthreshold interconnect performance in terms of speed and robustness has been suggested in Ref. [23]. However, interconnect analysis in the presence of process and temperature variations has not been taken into account. A subthreshold interconnect with a bootstrapped repeater scheme to accelerate the circuit operation and reduce the process variation has been proposed [24]. Significant research work has also been carried out on global interconnect design in super-threshold region [25–27]. However, less progress has been reported regarding interconnect performance for ultra-low-power VLSI applications. Since interconnects in VLSI circuits are driven by CMOS buffers, this includes the development of an analytical approach for performance investigation of buffer-driven long interconnects. SPICE simulations are best used for verification of the analytical techniques developed for VLSI analysis. Subsequently, analytical models if developed shall be particularly useful for ultra-low-power VLSI interconnect design. This is the primary focus of the book.

1.2 Motivation

A detailed literature survey has been carried out in the design of long interconnects for ultra-low-power VLSI chips and presented in Chap. 2 which clearly establishes that it is essential to investigate the interconnect circuit performance under subthreshold for ultra-low-power applications. The ubiquitous era of ultra-low-power applications such as microsensor networks, pacemakers, and many portable devices requires extremely energy stringent operation for longer battery lifetime. Subthreshold operation presents an opportunity for such emerging and energy-constrained applications in clock ranges from low to medium frequencies with very low energy consumption.

Extremely high driver resistance under subthreshold conditions and exponential dependence of subthreshold current on process parameters leads to an increased device susceptibility to variations in the process parameters. Furthermore, increased coupling among the adjacent interconnects has motivated the present research to investigate the performance of global interconnect under subthreshold conditions. To meet this end, the influence of the interconnect parasitics has been considered. Delay, power, crosstalk, and PVT variations of the long interconnects have been dealt with and investigated. This requires analytical modeling and simulation results. To meet out this requirement, problem has been tailored to fulfill the following objectives:

- To develop a timing model to characterize the output voltage and propagation delay of buffer-driven interconnect in subthreshold regime for ultra-low-power operation. It is also essential to consider the dependence of MOS, i.e., the active device behavior on input waveform.
- To develop analytical expressions characterizing the transient power dissipation of CMOS buffer-driven resistive–capacitive interconnect for subthreshold operation.
- To develop an analytical model to characterize the coupling noise in capacitively coupled interconnect scenario due to changing signal activity in subthreshold regime for ultra-low-power operation.
- To provide design guidelines for the reduction in coupling noise caused by the interconnect parasitics. Advantage of subthreshold interconnect circuit design has to be evaluated vis-a-vis super-threshold region.
- To investigate the effect of PVT variations on subthreshold interconnect performance for different interconnect and device parameters.

1.3 Book Outline

The main aim of the present work is to analyze and propose analytical models for buffer-loaded long interconnects in subthreshold for ultra-low-power VLSI circuits. This book is divided into six chapters. Main contributions and important findings are presented in the book in the following sequence:

- The present chapter introduces the topic and addresses the importance of interconnects vis-à-vis MOS devices. The chapter describes the motivation behind choosing investigation of interconnect performance in subthreshold as the research topic.
- Chapter 2 reviews the past research work carried out by various researchers in the design of digital subthreshold circuits to manage the proliferating demand of ultra-low-power devices. Various aspects of CMOS buffer have been studied. The role of interconnect parasitics on performance is examined. The detailed literature survey also deals with delay, power dissipation, and crosstalk in long interconnects. The effect of process and temperature variations on the interconnect performance has also been reviewed.
- Chapter 3 presents the timing analysis of a CMOS buffer stage resistive–capacitive interconnect in subthreshold regime. Analytical expressions characterizing the output voltage, propagation delay, and power dissipation have been presented. The analysis takes into consideration the dependence of MOSFET behavior on input waveform. Both rising and falling input ramps have been considered. Interconnect is modeled as resistive–capacitive load in order to emphasize the exponential behavior of CMOS logic gate in subthreshold. New sub-regions have been identified and their current–voltage equations also derived. Closed-form expressions of resistive power dissipation have also been presented.
- Chapter 4 deals with the evaluation of dynamic crosstalk due to the simultaneous switching of capacitively coupled interconnect. Analytical expressions for the output voltages and propagation delays of each CMOS inverter have been given in this chapter. These delay estimates are based on the assumption of fast ramp and slow ramp inputs. Variable load conditions and driver sizes of aggressor and victim have been considered.
- Chapter 5 characterizes the functional crosstalk at the output of the quiet driver due to signal activity on the aggressor driver. Analytical expressions characterizing undershoot and overshoot have been presented. Expressions governing the output voltage and propagation delay of aggressor driver have also been developed. Design guidelines to minimize the effects of coupling noise have been suggested. It is shown that subthreshold interconnect circuit design leads to significant savings in power-delay-crosstalk-product.
- Chapter 6 provides an in-depth analysis of the effect of PVT variations in sub-VLSI interconnects. In order to analyze the impact of PVT variations, Monte Carlo method and different process corners viz. fast–fast, slow–fast, typical–typical, fast–slow, and slow–slow have been used to analyze the effect on the performance of CMOS buffer-driven coupled interconnects.

Chapter 2
Design Challenges in Subthreshold Interconnect Circuits

Keywords CMOS buffer · Interconnect parasitics · Power dissipation · Weak inversion · Ultra-low-power

As technology advances to giga-scale integration level, global interconnect resource becomes increasingly valuable in a VLSI chip. This is due to the exponential growth of the total number of interconnects/wires as the feature size of MOS transistors decreases in scaled deep submicron CMOS technologies. Interconnect length, however, has not scaled down with feature size and remains long relative to other on-chip geometries. Interconnects are metal or polysilicon wires which connect billions of active devices to carry signals within a VLSI chip. There are a number of such wires in the whole chip. Of these, the length of long interconnects in large chips is of the order of 10 mm.

Interconnect and device performance in VLSI circuits depends on materials, geometry, and technology. With the dimensional scaling of technology, technological device and interconnect challenges have been closely examined by different researchers [28–33]. The delay in VLSI chips is due to active devices and interconnects. To avoid prohibitively larger delays, designers scale down global interconnect dimensions more slowly than the transistor dimensions [34]. Rather, reverse scaling is preferred for the global interconnects.

Interconnects also cause excessive power to be dissipated. In recent years, there has been a compelling demand for ultra-low-power devices to ensure longer battery lifetimes. Subthreshold circuits are ideally suited for applications where minimizing energy per operation is of prime importance [35–36]. Subsequently, the benefits from ultra-low energy operation have carved out a significant niche for subthreshold circuits. Furthermore, subthreshold circuits show exponential susceptibility to the process and temperature variations. Therefore, subthreshold operating region has made the design of energy-constrained robust ultra-low-power systems a very challenging design task. The present chapter reviews in detail the various aspects of buffer-driven long interconnect under subthreshold for ultra-low-power logic and the other associated problems. These are presented in the subsequent sections.

R. Dhiman and R. Chandel, *Compact Models and Performance Investigations for Subthreshold Interconnects*, Energy Systems in Electrical Engineering,
DOI 10.1007/978-81-322-2132-6_2

2.1 Interconnects for VLSI Applications—A Review

The long interconnects connect a larger number of active devices on a chip. These long interconnects distribute clock and signals and provide power and ground to the various circuits on a chip. Moreover, the associated parasitic impedance parameters increase as interconnect length increases. An overview of interconnect parasitics has been given in the following section.

2.1.1 Parasitic Impedance Parameters

Interconnect parasitics namely resistance, inductance, and capacitance lead to various undesirable effects in VLSI circuit design. These result in signal delay, power dissipation, distortion, and crosstalk. These problems are due to fundamental, material, device, circuit, and system physical limitations and need to be addressed while designing VLSI chips [37–39]. Interconnect has been represented by parasitic equivalent electrical components viz. resistance, inductance, and capacitance [40]. Such a lumped representation of the interconnect model is appropriate for medium and long interconnects at low-frequency applications. The parasitic impedance parameters are frequency dependent and responsible for decreased circuit efficiency and performance [41].

Eo and Eisenstadt [42] have developed models for high-speed and high-density VLSI circuit and found that interconnect circuit parameters vary with frequency. The model considers the silicon substrate properties, pad parasitics, fringing effects, and frequency variant properties of the circuit parameters. The model parameters are compared to scattering parameter measurements as well as PISCES-II simulations, and a good agreement is obtained with *s*-parameter measurements. Qian et al. [43] have developed an analytical expression for the effective load capacitance of resistive–capacitive (RC) interconnects. It is proved that, when there is a significant shielding, the response waveforms at the gate output may have a large exponential tail. This in turn can strongly influence the delay of RC interconnects. The concept of effective capacitance is extended to develop an equation on the basis of a two-piece gate-output approximation. The equation is solved accurately to obtain response waveform.

Delmore et al. [44], Moll et al. [45] and Wong et al. [46] have derived set of formulas to model capacitance and inductance in sub-half-micrometer VLSI circuits. Quasi three-dimensional (3D) modeling has been used for extracting the interconnect capacitance [47–48]. In the capacitance model, concept of effective width for a 3D wire has been used. This is derived from the combination of an analytical two-dimensional (2D) and 'wall-to-wall' model. The effective width provides a physics-based approach to decompose any 3D structure into a series of 2D segments, resulting in efficient and accurate capacitance extraction. Three-dimensional capacitance model for full-chip simulation has also been proposed in [49]. Huang et al. [50]

have carried out interconnect modeling for multi-gigahertz clock. However, accurate estimation of these interconnect parasitics requires details of the interconnect geometry, layout, technology, the current distributions and switching activities of the wires, which are difficult to predict and require more research. Rosa [51] has given the formula for self and mutual inductance using Biot–Savart law for linear conductors. Banerjee and Mehrotra [52] have introduced an accurate analysis of on-chip inductance effects for distributed interconnects that takes into account the effect of series resistance and output parasitic capacitance of the driver. The expressions for the transfer function of distributed interconnect lines, their time-domain responses, and computationally efficient performance optimization techniques have been presented. Closed-form approximation of frequency-dependent mutual impedance per unit length of lossy silicon substrate coplanar-strip IC interconnects has been developed in [53]. The derivation is based on a quasi-stationary full-wave analysis and Fourier integral transformation.

Sylvester and Hu [34] have considered the characterization of interconnect with particular attention to ultra-small capacitance measurement and in-situ noise evaluation techniques. An approach called the charge-based capacitance measurement technique, to measure Femto-Farad level wiring capacitances, has the advantages of being compact, having high-resolution and being very simple. Cong and Pan [54] have presented a set of interconnect performance estimation models for design planning with consideration of various effective interconnect layout optimization techniques. These models can be used efficiently during high-level design space exploration, interconnect-driven design planning, and synthesis- and timing-driven placement to ensure design convergence for deep sub-micrometer designs. A systematic method for deriving the characteristic model of interconnects from time-domain vector fitting has been investigated in [55]. The method is based on the iteration and convolution of time series by recursion. The approach extracts model parameters from terminal voltage waveforms directly by time-domain vector fitting so that the transformation of frequency loading can be simulated efficiently in SPICE-compatible simulator. The contributive interconnect parasitic impedance parameters contribute significantly to delay in VLSI chips. Estimation of propagation delay through interconnect has been of great concern for VLSI designers. Therefore, consideration of interconnect delay has been developed next.

2.1.2 Interconnect Delay

Interconnect delay modeling has been a subject of research since 1970s. Estimation of propagation delay through interconnect requires accurate models for the propagation path. Over the years, several models for interconnect delays have been proposed and tested. Resistive interconnect optimization under Elmore delay model [56] is carried out by Sapatnekar [57]. Gupta et al. [58] have proved that the Elmore delay measure is an absolute upper bound on the actual 50 % delay of RC tree response. Moreover, this bound holds for input signals other than steps. The actual delay asymptotically approaches the Elmore delay as the input signal rise time increases. A lower bound on delay is also developed using the Elmore delay and the second moment of the impulse response. Brocco et al. [59] have investigated macro-modeling and RC tree approaches giving a unified timing simulation method. The simulation method is faster than SPICE by two orders, for 2 μm CMOS technology. O'Brien and Savarino [60] have modeled the driving point characteristics of resistive interconnects for delay estimation. Compact expressions for worst-case time delay and crosstalk of coupled RC lines are proposed by Sakurai [61]. Kahng and Muddu [62] have developed an analytical delay model based on first and second moments to incorporate inductance effects in the delay estimation with step input. Delays estimated are within 15 % of SPICE-computed delay across a wide range of interconnect parameter values. A stochastic wiring distribution based upon Rent's Rule has been derived by Davis and Meindl [63–64]. The distribution determines wire-length frequency and enables a priori estimation of the local, semi-global, and global wiring requirements for future GSI systems. Brachtendorf and Laur [65] have provided analytical models by discretization of the telegrapher's equations, for the transient simulation of lossy interconnects. Chiprout [66] has presented guidelines for modeling on-chip interconnects for accurate simulation of high-performance ultra-large-scale integration designs. Pamunuwa and Tenhumen [67] have discussed the delay model for coupled interconnects. Analytical expressions for delay, buffer size, and number that are suitable in a priori timing analyses and signal integrity estimations have been developed.

Davis and Meindl [68, 69] have extended Sakurai's work [61] by including self and mutual inductance. The compact analytical expressions derived give an explanation for the transient response of high-speed distributed resistive–inductive–capacitive interconnect. Simplified expressions enable physical understanding and accurate estimation of transient response, propagation delay and crosstalk for global interconnects. Venkatesan et al. [70, 71] have significantly extended the work reported in Refs. [68, 69]. They have developed a new physical model for the transient response of distributed interconnects with a capacitive load. The solutions are verified by HSPICE simulations. These solutions are used to derive novel expressions for the propagation delay, optimum number, and size of buffers for buffer inserted distributed lines. The analysis defines a design space that reveals the trade-off between the number of buffers and wire cross section for specified delay and crosstalk constraints.

Xu and Mazumder [72] have introduced the passive discrete modeling technique using the numerical approximation method. This is called the differential quadrature method for estimating signal propagation delays through on-chip long interconnects. This delay modeling generates equivalent circuit interconnect models consisting of current and voltage sources, which can be directly incorporated into circuit simulators such as SPICE. Current sensing, model-reduction-based algorithms, etc., are some other delay analysis methods which have been proposed in [73]. Worst-case delay has been estimated by Chen et al. [74]. Singhal et al. [75] have presented a twofold approach for evaluating the signal and data carrying capacity of on-chip interconnects. In the first approach, the wire is modeled as a linear time invariant system and frequency response is studied and higher transmission rate is achieved using ideal signal shape. The second approach addresses delay and reliability in interconnects. Lehtonen et al. [76] have presented a self-contained adaptive system for detecting and bypassing permanent errors in on-chip interconnects. The proposed system reroutes data on erroneous links to a set of spare wires without interrupting the data flow. An improved syndrome storing-based detection method is presented and compared to the in-line test method. In the presence of permanent errors, the probability of correct transmission in the proposed systems is improved by up to 140 % over the standalone Hamming code. These methods achieve up to 38 % area, 64 % energy, and 61 % latency improvements at comparable error performance. Morgenshtein et al. [77] have presented a unified logical effort delay model for paths composed of CMOS logic gates and resistive wires. The method provides conditions for timing optimization while overcoming the limitations of standard logical effort in the presence of interconnects. The condition of optimal gate sizing in a logic path with long wires is also given and the condition is achieved when the delay component due to the gate input capacitance is equal to the delay component due to the effective output resistance of the gate.

2.1.3 CMOS Buffer

It is an important technique in VLSI to drive interconnects by buffers. Buffers have been realized using CMOS inverters [41]. Researchers have modeled buffers differently and much work has been reported in literature about CMOS buffers. Shockley [78] and Shichman and Hodges [79] have developed square law models for MOSFETs in which drain current varies as a square of the effective gate voltage. These models have been extensively used in computer-aided analyses of CMOS switching circuits. However, the proposed models do not give accurate results as channel length is reduced. Sakurai and Newton [80] developed alpha-power model which defines current–voltage characteristics for short-channel transistors. In this model, the input waveform slope effects and parasitic drain/source resistance effects are included. It has been observed that neglecting *p*-channel transistor (PMOS) is not valid when the input ramp is very slow compared to the output waveform. However, the approximation is valid if the input slope exceeds one-third of the output slope, which is usually true in VLSI. Various approaches for taking into consideration the non-ideal effects in short-channel MOSFETs have been considered [80].

Deng and Shiau [81] have used the linear RC delay method to empirically calculate the delay in digital CMOS circuits. This generic linear RC model has the advantage of being simple and reliable. The empirical model, which is a high-dimensional function of various circuit and device parameters, is simplified to a 2D model that estimates the delay of CMOS circuits. SPICE simulation is used to verify the analytical results. Chung et al. [82] have carried out a comprehensive study of the performance and reliability design issues for deep submicrometer

MOSFET. The performance criteria viz. current-driving capability, ring-oscillator switching speed, and small-signal voltage gain are studied. In this context, the allowable choice of MOSFET channel length, oxide thickness, and power supply voltage is examined. Dutta et al. [83] have developed an analytical and comprehensive scheme to evaluate the delay and the output transition time of buffer for any input ramp and different fan-outs. Turn points for the infinitely fast and infinitely slow input rise times have been identified. A smooth curve fitting is used to predict the delay and the transition time over a large range of input signal slopes and output loading. The accuracy of the analysis is within 3 % of SPICE results. Bisdounis et al. [84] have suggested analytical transient and propagation delay models for short-channel CMOS inverter with fast and slow ramp inputs. They have used alpha-power law MOSFET model and taken gate-to-drain coupling capacitance into account. The analytical results show an error less than 3 %. The reduction of transistor-level models of CMOS logic gates to equivalent inverters, for the purpose of computing the supply current, power and delay in digital circuits has been carried out by Nabavi-Lishi and Rumin [85]. Hirata et al. [86] have derived propagation delay for static CMOS gates considering short-circuit current and the currents through capacitive load and gate capacitance. They demonstrated that the influence of short-circuit power on delay becomes large with slow input transition and small output load capacitance. The accuracy of this analytical method is better than that reported in [80], especially when the velocity saturation is large. The error of the analysis is within 8 % of SPICE results. Pattanaik et al. [87] have used geometric programming for the optimization of delay and power of nanoscale CMOS inverter.

Daga and Auvergne [88] have demonstrated a design-oriented comprehensive analytical model for CMOS inverter delay considering input slope, input-to-output capacitive coupling, short-circuit current, and short-channel effects. Gate input dependency and the input-slope-induced nonlinearity are considered. The overall calculated results are within 10 % of SPICE simulation results. Raja et al. [89] have given a new CMOS gate design that has different delays along various inputs to output paths. The delays are accomplished by inserting selectively sized permanently on series transistors at the inputs of a logic gate. The use of variable input delay CMOS gates for total glitch-free minimum dynamic power implementations of digital circuits has been demonstrated. Using c7552 benchmark circuit and described gates, power saving of 58 % is obtained. CMOS gate sizing, taking the dependence of fan-out, spurious capacitances and the slope of the input waveforms to optimize delay has been presented [90]. The alpha-power law model equations have been used.

2.2 Coupling Capacitance Noise

Interconnect coupling noise or crosstalk refers to the voltage induced on the victim node due to capacitive coupling from a switching aggressor node. The coupling capacitance causes disastrous effect on the functionality and reliability of digital integrated circuits. It induces a voltage glitch in one or more adjacent quiet interconnects and may become a cause for circuit failure [91]. It also leads to false propagation delay times and increased power dissipation. In modern interconnect design, interconnects in adjacent metal layers are kept orthogonal to each other. This is done to reduce crosstalk as far as possible. But with growing interconnect density and reduced chip size, even the non-adjacent interconnects exhibit coupling.

The extent of coupling is dependent upon the nature of the signal transitions [92–94]. If both interconnects switch in the same direction, the coupling capacitance (C_c) is approximately zero and the total capacitance of each interconnect is approximated by the line-to-ground capacitance. If one interconnect is switching and the other is quiet, the total capacitance of each interconnect is determined by the capacitance ($C + C_c$). On the other hand, if the signals on each interconnect switch out of phase, the effective coupling capacitance approximately doubles to $2 \times C_c$. Thus, the coupling capacitance changes the effective load capacitance, depending upon the signal switching activity. Buffer insertion is a technique commonly used for the reduction of crosstalk. However, the buffer insertion technique in subthreshold is not a feasible technique contrary to the super-threshold region. Another useful approach of reducing crosstalk is to use shielding wires, which also increases the capacitive load and therefore delay. A more suitable approach is to increase the spacing between the wires.

Extensive research has been carried out regarding crosstalk and delay estimation of CMOS gate-driven coupled interconnects. Xie and Nakhla [95] have proposed a method for crosstalk and delay estimation in high-speed VLSI interconnects with nonlinear components. The solution of the mixed frequency and time-domain problem by replacing the linear subnetworks, with a set of ordinary differential equations using the asymptotic waveform evaluation technique, has been obtained. Poltz [96] has given electromagnetic modeling of VLSI interconnects and the Helmholtz equation is used to build models which include eddy current loss and dielectric loss. Equivalent circuits with high cutoff frequencies and the smallest possible number of components are assembled. The performance of a VLSI interconnect at different clock rates is analyzed. Kuhlmann et al. [97] have proposed a time-efficient method for the precise estimation of crosstalk noise. A metric to compute coupling noise according to the sink capacitances and conductances of the aggressor and victim nets has been reported. The noise waveform is computed using a closed form leading to short computation time. The problem of crosstalk computation and reduction using circuit and layout techniques has been addressed in [98–99]. Expressions have been provided for noise amplitude and pulse width in capacitively coupled resistive lines. The estimation is based upon the RC transmission line model. A three-line structure of coupled RC interconnects using the transmission line model is presented in [100]. However, MOS transistor has been approximated by a linear resistor. Ling et al. [101] have developed a method to estimate the coupling noise in the presence of multiple aggressor nets. Authors have reported a novel technique for modeling quiet aggressor nets based on the concept of coupling point admittance and a reduction method to replace tree branches with effective capacitors. The proposed method has been tested for noise-prone interconnects from an industrial high-performance processor in 0.15 μm technology.

The worst-case error of 7.8 % and an average error of 2.7 % are observed. Devgan [102] has presented a metric for estimation of coupled noise in on-chip interconnects. This noise estimation metric is an upper bound known as the Devgan metric for RC circuits, being similar in spirit to Elmore delay in timing analysis. An enhancement to the Devgan metric has been proposed in [103] to improve the accuracy for fast input signals. The coupling noise voltage on a quiet interconnect line has also been analyzed by Shoji using a simple linear RC circuit [104]. Hashimoto et al. [105] have proposed a method to capture crosstalk-induced noisy waveform for crosstalk aware static timing analysis. The static timing analysis is performed with the consideration of dynamic delay variation due to crosstalk noise. Eo et al. [106] have proposed a simple closed-form crosstalk model and experimentally verified the model with 0.35 μm CMOS process-based interconnect test structures having two, three and five coupled lines with different switching scenarios. Becer et al. [107] have presented a complete crosstalk noise model which incorporates all victim and aggressor driver/interconnect physical parameters including coupling locations on both victim and aggressor nets. The validity of given model against SPICE has been demonstrated and has a good trade-off between accuracy and completeness, having an average error of 10 % with respect to SPICE for 130 nm technology. Hasan et al. [108] have derived and analyzed the crosstalk noise effect on a single victim line. An accurate and flexible decoupled transient model for victim wire is introduced. The model can be used to compute the maximum delay and glitch effect due to crosstalk under different slew rates. Tuuna et al. [109] have given an analytical model for the current drawn by on-chip bus. The model is combined with an on-chip power supply grid model in order to analyze noise caused by switching buses in a power supply grid. The buses are modeled as distributed lines that are capacitively and inductively coupled to each other. Different switching patterns and driver skewing times are also included in the model.

Bazargan-Sabet and Renault [110] have presented closed-form formulas to estimate capacitive coupling-induced crosstalk noise for distributed RC coupling trees. The formulas are simple enough to be used in the inner loops of performance optimization algorithms or as cost functions to guide routers. Kaushik et al. [111] have considered the effect of crosstalk-induced overshoot and undershoot generated at noise-site. The false switching occurs when the magnitude of overshoot or undershoot is beyond the threshold voltage of the gate. The peak overshoot and undershoot generated at noise-site can wear out the thin gate oxide layer resulting in permanent failure of the VLSI chip. Agarwal et al. [112] have analyzed a simple crosstalk noise model for coupled on-chip interconnects. The model is based on coupled-transmission-line theory and is applicable to asymmetric driver and line configurations. The noise waveform shape is captured well and yields an average error of 6.5 % for noise peak over a wide range of test cases. Chen and Sadowska [113] have proposed closed-form formula to estimate capacitive coupling-induced crosstalk noise for distributed RC coupling trees. The efficiency of the approach stems from the fact that only the five basic operations are used in the expressions viz. addition, subtraction, multiplication, division, and square root. Lee et al. [114]

have given crosstalk estimation method using coupled inductive tree models in high-speed VLSI interconnect. The recursive formulas for moment computation of coupled inductive interconnect trees with self and mutual inductances have been generalized. Nieuwoudt et al. [115] have given a comprehensive investigation of crosstalk-induced delay, noise, and capacitance for 65 nm process technology. Naeemi et al. [116] have described an analytical model that describes distributed inductive interconnects with ideal and non-ideal return path to optimize crosstalk and time delay of high-speed global interconnect structures such that the crosstalk and delay reduce by 38 and 12 %, respectively.

Vittal et al. [117] have addressed the problem of crosstalk computation and reduction using circuit and layout techniques. The expressions for crosstalk amplitude and pulse width in capacitively coupled resistive lines have been provided. The expressions hold good for nets with arbitrary number of pins and of arbitrary topology under any specified input excitation. The experimental results show that the average error is about 10 % and the maximum error is less than 20 %. Avinash et al. [118] have proposed a spatiotemporal bus encoding scheme to minimize crosstalk in interconnects. The scheme eliminates crosstalk in the interconnect wires, thereby reducing delay and energy consumption. The technique is evaluated by focusing on L1 cache address/data bus of a microprocessor using SPEC2000 CINT benchmark and suites for 90 and 65 nm technologies.

Nuroska et al. [119] have given a technique that reduces crosstalk noise on buses based on profiling the switching behavior. Based on this profiling information, an architecture configuration obtained using a genetic algorithm is applied that encodes pairs of bus wires, permutes the wires, and assigns an inversion level to each wire in order to optimize for noise and power. Hanchate and Ranganathan [120] have proposed a methodology for wire sizing with simultaneous optimization of interconnect crosstalk noise and delay in deep submicron VLSI circuits. The wire sizing is modeled as an optimization problem, formulated as a normal form game, and solved using the Nash equilibrium. Game theory allows the optimization of multiple metrics with conflicting objectives. Lienig [121] presented a novel approach to solve the VLSI channel and routing problems. The approach is based on a parallel genetic algorithm which runs on a distributed network of workstations. The algorithm optimizes physical constraints such as the length of nets, number of vias and is able to significantly reduce the occurrence of crosstalk.

Rao et al. [122] have proposed a bus encoding algorithm and circuit scheme for on-chip buses that eliminates capacitive crosstalk while simultaneously reducing total power. The encoding scheme significantly reduces total power by 26 % and runtime leakage power by 42 % while eliminating capacitive crosstalk. Zhang and Sapatnekar [123] have presented a method for incorporating crosstalk reduction criteria into the global routing under a broad power supply network paradigm. The method utilizes power/ground wires as shields between the signal wires to reduce capacitive coupling, while considering the constraints imposed by limited routing and buffering resources. An iterative procedure is employed to route signal wires, assign supply shields, and insert buffers. Wu et al. [124] have proposed a probabilistic model-based approach for crosstalk mitigation at the layer assignment. The approach aims to discover and reduce crosstalk at the pre-detailed-routing level. Ho et al. [125] have given a novel framework for fast multilevel routing considering crosstalk and performance optimization. An intermediate stage of layer/track assignment has been incorporated into the multilevel routing framework. Compared with the state-of-the-art multilevel routing, the experimental results show that their approach achieved a 6.7× runtime speedup, reduced respective maximum and average crosstalk by about 30 and 24 %, and reduced respective maximum and average delay by about 15 and 5 %.

Yoshikawa and Terai [126] have examined crosstalk-driven placement procedure based on genetic algorithm. For selection control, objective functions are introduced for improving crosstalk noise, reducing power consumption, improving interconnection delay, and dispersing wire congestion. Authors in [127] have proposed a coupling-driven data encoding scheme for low-power data transmission in deep submicron buses. The encoding scheme reduces the coupling transitions by 23 % for a deep submicron bus compared to the non-coded data transmission. It has been found that 75 % of the power consumption is due to coupling capacitance, whereas 25 % is due to self capacitance.

2.3 Power Dissipation

With the emergence of portable computing and communication equipments, low-power design has become a principal theme of the VLSI industry. The need for portability has caused a major paradigm shift in which power dissipation is as important as speed and area. The most demanding applications of low-power microelectronics have been battery-operated wrist watches, hearing aids, implantable cardiac pacemakers (a few μW power consumption), pocket calculators, pagers, cellular telephones (a few mW), and prospectively the hand-held multimedia terminals (10–20 W). The power dissipation in VLSI circuits is reviewed in this section. Various methodologies for reduction of power dissipation in VLSI circuits are also examined.

Powers [128] discussed the existing and emerging battery systems in terms of energy content, shelf and cycle life besides other characteristics. Progress in battery technology is still far behind than that in the field of electronics. Packaging has resulted in significant changes in the older systems such as C–Zn, alkaline, Zn–Air, NiCd, and lead acid which continue to get better. Chandrakasan et al. [129] have presented an analysis of low-power CMOS digital design, giving the techniques for low-power operation that use the lowest possible supply voltage coupled with architectural, logic style, circuit, and technology optimizations. The optimum voltage for 2 μm technology is 1.5 V and for 0.8 μm technology is 1 V, with power dissipation reduction by a factor of 10. The architectural-based scaling strategy indicates that the optimum voltage is much lower than that determined by other scaling considerations. Davari et al. [130] have given guidelines of CMOS scaling for low-power design. Comparisons are given for CMOS technologies ranging from

0.25 μm at 2.5 V to sub-0.1 μm at 1 V. It is shown that over two orders of magnitude improvement in power-delay-product are expected by such scaling compared to 0.6 μm devices at 5 V supply. Meindl [131] meticulously discussed the pros and cons of future opportunities for low-power GSI which are governed by a hierarchy of (i) theoretical and practical, (ii) material, (iii) device, (iv) circuit, and (v) system limits.

Low power is an essential requirement of biomedical electronic devices. Bhattacharyya et al. [132] have developed low-power hearing aid circuit based on 1 V supply voltage and adaptive biasing. Corbishley et al. [133] have proposed an ultra-low-power analog system to provide adaptive directionality in digital hearing aids. Power reduction is obtained by designing all the circuit blocks, viz. filters, multipliers, and dividers, in CMOS technology using transistors in weak inversion region. The total power consumption of the complete system is 5 μW at a scaled supply voltage of 0.9 V in 0.35 μm technology. Various power estimation techniques have been surveyed by Najm [134]. Rajput and Jamuar [135, 136] have reported low-voltage analog VLSI circuit design techniques and their applications. Power dissipation analysis of DSM CMOS circuits is carried out by Gu and Elmasry [137]. Borah et al. [138] and Heulser and Fichtner [139] considered transistor sizing for minimizing power consumption of CMOS circuit under delay constraint.

Authors in [140–143] have described several methodologies for low-power VLSI design. To contain the adverse effects of power dissipation, low-voltage operation of circuits, along with variable threshold and multiple threshold CMOS techniques, is often resorted to. System level architectural measures such as pipelining approach and parallel processing or hardware replication technique are used in the trade-off areas for low-power dissipation. Reduction of switching activity by algorithmic, architectural and circuit level optimization by proper choice of logic topology reduces power dissipation. Delay balancing, glitch reduction, and use of conditional or gated clock signals are some of the useful architectural measures to reduce switching activity. Switched capacitances play a significant role in switching power dissipation. Reduction of switched capacitances is a major step for low-power design of digital ICs. This can be accomplished (i) at system level by limiting the use of shared resources, e.g., by partitioning the global bus into smaller dedicated local buses to handle data transmission between nearby modules, (ii) by using proper logic style e.g., pass transistor logic reduces load capacitance, and (iii) by reducing parasitic capacitance at physical design level by keeping transistors at minimum dimensions whenever possible.

Kang [144] has reported an accurate method for simulating the power dissipation in an IC by the use of a dependent current source and a parallel RC circuit. The steady-state voltage across the capacitor reads the average power drawn from the supply voltage source. Simulation results are shown for CMOS circuits. This subcircuit can be inserted into any VLSI circuit model without causing interference while the circuit is simulated with a simulator such as SPICE. Yacoub and Ku [145] envisioned a circuit simulation technique which permits the measurement of short-circuit power dissipation component in ICs using SPICE. This technique is most appropriate for low-power CMOS circuit design that does not permit current flow, other than leakage current, during steady-state operation.

Constandinou et al. [146] implemented an ultra-low-power consuming, simple, and robust circuit for edge-detection in integrated vision systems in 0.18 μm CMOS technology. Kim et al. [147] presented a low-power smallest area, delay-locked loop-based clock generator. Fabricated in a 0.35 μm CMOS process, clock generator occupies 0.07 mm^2 area and consumes 42.9 mW power and operates in the frequency range of 120 MHz–1.1 GHz. Bhaumik et al. [148] implemented a divided word line scheme to bring down power dissipation in 256 kB static random access memory design. Mitra and Chandorkar [149] designed a low-voltage CMOS amplifier with rail-to-rail input common mode range. Alternative methods were applied for obtaining high common mode range, good common mode rejection ratio, and output swing at such low supply voltage. Hwang et al. [150] reported a self regulating CMOS voltage-controlled oscillator with low supply voltage sensitivity. Lidow et al. [151] examined future trends in Internet appliances, portable electronic appliances, and silicon-based power transistors and diodes. It is discussed how the changing requirements of end users are driving state-of-the-art devices, new analog ICs as well as different power management architectures. Methodologies and projections related to power dissipation in CMOS circuits have been specified by Bhavnagarwala et al. [152].

Mutoh et al. [153] have proposed circuit by inserting high-threshold devices in series into low-threshold circuitry. A sleep control scheme is introduced for efficient power management. Kawaguchi et al. [154] have suggested super cutoff CMOS circuit that uses low-threshold voltage transistor with an inserted gate bias generator. In the standby mode, the voltages are applied to transistors to fully cut off the leakage current. Wei et al. [155] have implemented the dual-threshold technique to reduce leakage power by assigning a high-threshold voltage to some transistors in non-critical paths and using low-threshold transistors in the critical path. An algorithm for selecting and assigning an optimal high-threshold voltage is also given. The reduction in leakage power is more than 80 % and total active power saving is around 50 and 20 %, respectively, at low- and high-switching activities for ISCAS benchmark circuits. In [156], the authors have presented architectures for low power and optimum speed for image segmentation using Sobel operators.

Pant et al. [157] have presented algorithms that can be used to design ultra-low-power CMOS logic circuits by joint optimization of supply voltage, threshold voltage, and device widths. Various components of power dissipation are considered and an efficient heuristic is developed that delivers over an order of magnitude savings in power over conventional optimization methods. The authors have also proposed a heuristic technique for minimizing the total power consumption under a given delay constraint. The approach simultaneously determines transistor power supply, threshold voltage, and device width by two distinct phases. The proposed approaches trade off energy and delay invariably by tuning variables (supply voltage, threshold voltage, transistor size, etc.). Chi et al. [158] have proposed a multiple supply voltage-scaling algorithm for low-power design. The algorithm combines a greedy approach and an iterative improvement optimization approach.

Deodhar and Davis [159] have suggested voltage-scaling and repeater insertion for throughput-centric low-power global interconnects. It is assessed that repeater insertion improves throughput. Using 180 nm technology, it is illustrated that 1 V supply voltage can reduce power dissipation up to 25 % of that with 2.5 V supply, for 2 Gbps throughput. The results are compared with SPICE simulations and show a good agreement. The possibility of applying the buffer insertion technique to reduce power dissipation and delay in interconnects in voltage-scaled environment has been carried out in [160, 161]. Analytical approaches for optimum design and optimum number of buffers in low-power environment have been developed. Buffer sizing for minimum power and delay in voltage-scaled environment has also been carried out. The analytical results are within 10 % of the SPICE simulation results. Banerjee and Mehrotra [162] have addressed the problem of power dissipation during the buffer insertion phase of interconnect design optimization. Since all global interconnects are not on the critical path, a small delay penalty can be tolerated on these non-critical interconnects. A delay penalty of 5 % for lesser power dissipation at different MOS technologies has been included. It is proved that there exists a potential for large power saving by using smaller buffers and larger inter-buffer interconnect lengths. Wang et al. [163] have represented signals by localized wave packets that propagate along the interconnect lines at the speed of light to trigger the receivers. Energy consumption is reduced through charging up only part of the interconnect lines. Voltage doubling property of the receiver gate capacitances is used. Zhong and Jha [164] demonstrated the importance of optimizing on-chip interconnects for power reduction. It is concluded that significant spurious switching activity occurs in interconnects.

Tajalli and Leblebici [165] experimentally and analytically showed that scaling supply voltage in deep subthreshold region increases energy consumption and also investigated that optimum supply voltage for minimum energy consumption lies in moderate subthreshold region. Moreover, digital circuits operated in deep subthreshold region will have significant delay and noise margin penalties along with robustness issues that cannot be ignored for portable devices with real-time applications [166]. Hence, the designing of robust and moderate performance subthreshold field programmable gate arrays, real-time portable devices, buses, and clock signal is uncertain at such low bias [167].

Bol et al. [168] investigated the interests and limitations of technology scaling for subthreshold logic from 0.25 μm to 32 nm nodes. It is shown that scaling from 90 to 65 nm nodes is highly desirable for medium-throughput applications (1–10 MHz) due to great dynamic energy reduction. Upsizing of the channel length as a circuit level technique has been proposed to efficiently mitigate short-channel effects.

Thus, from the literature, it is clear that reducing power dissipation has been a crucial parameter for low-power VLSI designs. Also, energy-constrained VLSI applications have emerged for which the energy consumption is the key metric and speed of operation less relevant. The power consumption of these systems should decrease to the extent so as to extend the battery life and theoretically have unlimited lifetimes [169]. To cope with such ultra-low-power applications, design of digital subthreshold logic was investigated. In the next section, fundamental aspects of subthreshold design for ultra-low-power circuits have been provided.

2.4 Weak Inversion for Ultra-Low-Power Logic

When gate-to-source voltage is less than or equal to the transistor threshold voltage, transistor is said to be biased in subthreshold. The transistor conducts current through an inverted channel between the source and drain caused by a voltage applied to the gate. The majority carriers in the substrate are repelled from the surface directly below the gate. A depletion charge of immobile atoms forms a depletion layer beneath the gate. The minority carriers in the depletion layer are made to move by diffusion and induce a drain current when a voltage that is less than the threshold voltage is applied between the drain and source terminals in the MOS device. This current is referred to as weak inversion current or subthreshold current. Due to small drive current, the subthreshold logic only fits in designs, where the performance is secondary and not the main concern. Since the leakage current is orders of magnitude lower than the drain strong inversion current and the power supply is reduced, subthreshold logic dissipates ultra-low-power [12, 13, 170]. The subthreshold circuit designs therefore offer significant savings in energy because reduction in power consumption outweighs the increase in delay by an order of magnitude [171]. These also provide near ideal voltage characteristic curve, a requirement for digital circuits [172]. Furthermore, in the subthreshold region, the transistor input capacitance is lesser than its strong inversion counterpart [36]. The low-operating frequency, low supply voltage, and smaller input gate capacitance combine together to reduce both dynamic and leakage power. A number of other advantages in subthreshold operation include improved gain, better noise margin, and tolerant to higher stack of series transistors.

Subthreshold digital operation was first examined theoretically in 1970s in the context of studying the limits of voltage scaling [173]. Subthreshold design was explored for low-power analog applications such as amplitude detector, quartz ring oscillator, band pass amplifier, and transconductance amplifier [174–176]. Digital subthreshold circuits were slower to catch on.

In the past years, a growing number of successful implementations of digital subthreshold systems viz. biomedical devices, fast fourier transform (FFT) processors, sensors, and static random access memory (SRAM) [18, 177–181] have occurred. An ultra-low-power delayed least mean square adaptive filter for hearing aids that uses parallelism is reported in [14]. The adaptive filter achieves 91 % improvement in power compared with a non-parallel CMOS implementation. The filter gives the desired performance of 22 kHz and operates at 400 mV. In 2001, Paul et al. designed an 8 × 8 array multiplier in 0.35 μm technology to operate in subthreshold operation [182]. The power-delay-product of this multiplier is around 25 times lower than its strong inversion operation. Body biasing is used to reduce the multiplier delay occurring due to temperature changes. A 2.60 pJ/instruction subthreshold sensor processor in 0.13 μm technology has been fabricated in [183]. The minimum energy consumption is improved 10 times that of the previous sensor processor. A 180 mV subthreshold processor using FFT in 0.18 μm technology has been fabricated by Wang and Chandrakasan [16, 184]. The FFT processor dissipates 155 nJ for 16 bits and 1,024 point FFT at the optimum supply voltage. Besides the subthreshold static logic, other logic families such as subthreshold pseudo-NMOS, variable threshold voltage subthreshold CMOS, subthreshold dynamic threshold voltage MOS, and subthreshold dynamic logic have also been proposed [185, 186].

Thus, ultra-low-power applications have established a significant niche for subthreshold circuits [11]. In future CMOS technologies, domination of subthreshold logic over super-threshold logic for ultra-low-power moderate throughput applications is expected. However, process and temperature variations have become one of the most challenging obstacles in subthreshold circuits in recent deep submicron technologies. The process variations are dramatically accentuated in subthreshold designs. This topic is addressed in the following section.

2.5 Variability in Subthreshold Design

The variation occurring in the various design parameters of transistor viz. threshold voltage, oxide thickness, channel length, and mobility during the IC fabrication is termed as process variation. It may also be defined as fluctuations around the desired value of design parameters introduced during chip device fabrication [187]. The process variation issue is important in present day IC design [4]. This section briefly introduces this topic and thereafter, a survey of the literature that addresses variability in subthreshold circuits is conducted.

The impact of process variations on power and timing has become significant especially beyond 90 nm since the fabrication process tolerances have not scaled proportionally with miniaturization of the device dimensions [21]. As CMOS devices are further scaled in the nanometer regime, variations in the number and placement of dopant atoms in the channel region, called random dopant fluctuation (RDF), cause random variations in the threshold voltage. RDF makes it increasingly difficult to achieve threshold voltage accuracy. This further exaggerates the variability problem by producing variations in the subthreshold swing, drain current, and subthreshold leakage current [188]. Shockley [189] during his research on random fluctuations in junction breakdown first discovered random variation phenomenon in semiconductor devices. He explained that variations in the threshold voltage are randomly distributed according to Poisson distribution. Keyes further extended Shockley's work by studying the effect of variability on the electrical characteristics of a MOSFET [190]. These variations cause different relative strengths of the constituent transistors, thereby causing functional failure of logic gates [191]. Consequently, the output voltage rise and fall times differ, thus impacting the switching frequency or power consumption. Body bias compensation circuits have been used to mitigate mismatch [192].

Authors in [193] have reduced the sensitivity to RDF through circuit sizing and the choice of circuit logic depth. The statistical models for circuit delay, power, and energy efficiency have also been derived. Kim et al. [194] have reduced the impact of RDF by device sizing optimization process which uses the reverse short-channel effect present in standard CMOS non-uniform halo doping profile devices. A transistor-level yield optimization technique to suppress process-induced variability has been proposed in [195].

Besides process variations, temperature variation has also a significant impact on the performance of subthreshold systems. The sources of temperature variations in VLSI circuits include ambient temperature and self heating. An increase in temperature increases the subthreshold leakage current and leakage power. This leakage power can be several orders of magnitude at higher temperatures.

2.5.1 Process Variations

The subthreshold current is exponentially dependent on the transistor threshold voltage. The threshold voltage is strongly related to the various device parameters such as effective gate length, oxide thickness, and doping concentration. These device parameters vary considerably in the DSM regime [196]. For example, a 10 % variation in the transistor effective length can lead to as much as a threefold difference in the amount of subthreshold leakage current [197]. Thus, subthreshold circuit designs are prone to process variations since this current drives the circuits.

The process variations have been classified into random and systematic variations [198]. Random variations can cause a device mismatch of identical and adjacent devices. The mismatch in the threshold voltage caused due to random variations decreases with decrease in doping and gate oxide thickness and increases when effective gate length and width decrease. This has been shown by Stolk et al. [199]. Systematic variations have been classified into across-field and layout-dependent variations. A cross-field variations cause identical devices in different parts of the chips to behave differently. Layout-dependent variations cause different layouts of the same device to have different characteristics. Even for the chips that meet the required operating frequency, a large portion dissipates very large amount of leakage power. This makes ICs unsuitable for commercial use [200]. These errors are caused due to photolithographic and etching sources, lens aberrations, mask errors, and variations in etch loading [201, 202]. Authors in [203] have carried out Monte Carlo analysis for a small MOSFET. They showed that controlling the process variation parameters to ±10 % yields a threshold voltage variation of ±15 %. They also showed that 95 % of the variance was around ±100 mV about the mean threshold voltage with normal distribution. Bauer et al. [204] have shown that threshold voltage depends upon the depth of penetration of ions during ion implantation. Schemmert and Zimmer [205] introduced a procedure for minimizing this threshold voltage sensitivity of ion-implanted MOSFETs. Their results showed a maximum deviation of 10 %. Kuhn et al. [206] have demonstrated that process variations also affect high-dielectric metal gates resulting in parametric variations in drive current, gate tunneling current, and threshold voltage. Recent work on process variation in subthreshold has focused upon the design of ultra-low-voltage SRAM with techniques reported for improving robustness [17, 207]. However, it fails to operate in the presence of process variability. Zhai et al. have highlighted three design challenges to ultra-low-voltage subthreshold SRAM [178]. These design challenges are (i) increased sensitivity to process variations, (ii) reduced on-current to off-current ratio which leads to difficulty in distinguishing between the read current of an accessed cell and the leakage current in the unaccessed cell, and (iii) the change in gate sizing requirements. The read and write stabilities of SRAM are heavily dependent upon the pull-up, pull-down, and pass transistors. The authors have presented a six-transistor SRAM design in subthreshold capable of overcoming aforementioned design challenges. The proposed design provides 36 % improvement in energy over other proposed SRAM designs with less area overhead.

Thus, process variation plays a key role in deciding robustness and energy efficiency of subthreshold designs. CMOS literature has always shown process variation as a critical element in semiconductor fabrication. The next section conveys some considerations regarding the effect of temperature in subthreshold VLSI circuits.

2.5.2 Temperature Variations

The temperature effects on subthreshold circuit operation have been investigated by Datta and Burleson [208]. It has been found that current exhibits positive temperature coefficient and increases exponentially with temperature while the on-to-off current ratio degrades by 0.52 %/°C due to the relative increase in leakage currents. Effect of temperature on subthreshold interconnect performance has also been carried out. It is found that optimal energy-delay-product can be achieved in the high temperature range of 75–90 °C. Authors in [209] have proposed an architecture and circuit for temperature sensors in 0.8 μm CMOS technology for ultra-low-power applications. The sensor draws 40 nA current from 1.6 to 3 V supply at room temperature. The circuit is suitable for temperature sensing in the 290–350 K temperature range. A temperature sensor suitable for passive wireless applications has been fabricated in 0.18 μm CMOS technology [210]. The temperature sensor consumes only 220 nW at 1 V at room temperature. It exhibits temperature inaccuracy of −1.6 °C/+3 °C from 0 to 100 °C. Hanson et al. [211] have designed a processor for sensor applications. The processor is capable of working at 350 mV operating voltage in the subthreshold region while consuming only 3.5 pJ of energy per cycle.

The effect of temperature on the various leakage current components has also been explained by the various researchers. The thermal dependence of punch-through current has been explained in [212]. It is found that punch-through current reduces at low temperatures. The temperature dependence of drain-induced barrier lowering (DIBL) current in deep submicron MOSFETs has been extensively investigated in [213]. It is found that the DIBL coefficient is nearly insensitive to temperature reduction in the temperature interval from 300 to 50 K. Authors in [214] show that DIBL coefficient increases nearly 2.5 times under temperature reduction from 150 to 25 °C. The dependence of impact ionization current component has been studied in [215–217]. The leakage current due to impact ionization is temperature independent in the temperature interval from 300 to 77 K and significantly increases under technology scaling.

2.6 Concluding Remarks

From the literature survey, it is observed that parasitics are associated with VLSI global interconnects, which hamper the performance of integrated circuits. Significant research has been carried out on optimal interconnect design in superthreshold region. However, very limited literature deals with interconnect design challenges under subthreshold conditions. Further study of interconnect design, for ultra-low-power environment, is needed. Increased delay and crosstalk have become challenging design issues particularly for subthreshold interconnects. The driver delay rather than the interconnect delay dominates under subthreshold conditions. Subthreshold attracted attention in the digital domain in the late 1990s. Since then, several subthreshold systems have been implemented with standard deep submicron technologies. Subthreshold circuits have been best suited to meet the growing demand for battery-operated portable ultra-low-power VLSI applications. Process, voltage, and temperature variations on the subthreshold circuit behavior have been analyzed. It is observed that variability is accentuated in subthreshold designs and is one of the most challenging obstacles in deep submicron technologies. From the literature review, it is thus concluded that buffer-driven interconnects under subthreshold need investigation as these are useful for energy-constrained ultra-low-power applications. Alternatively, there are more precise existing circuit MOS models such as EKV, BSIM, and high-level empirical models implemented in HSPICE for the evaluation of CMOS circuit performance. These models do not provide a closed-form expression for the characteristics of the MOSFET. However, the relationship between the geometric structure and the electrical behavior can be elucidated appropriately only by compact analytical techniques.

Chapter 3
Subthreshold Interconnect Circuit Design

Keywords Drain conductance · Propagation delay · Resistive power · Sub-saturation · Sub-linear

With the advancement in integrated circuit technologies, feature size of the MOS transistors has been aggressively scaled down with a remarkable increase in resulting integration density and chip size. The trend toward larger chip size has necessitated using longer interconnects. These connect various components on a very-large-scale integration chip and distribute power, ground, clock, data, and control signals. The performance of a logic gate is dictated by interconnects particularly in deep submicron technologies. The associated impedance parasitics degrade the overall performance significantly. Interconnect parasitic capacitance presents considerable loading to CMOS circuits, thereby increasing the propagation delays. These also cause excess power to be dissipated. Global or long interconnects in nanometer technologies have attracted increasing attention because of their growing influence on the overall performance of integrated circuits over the past few years.

In this chapter, an analytical framework to characterize the output voltage and propagation delay of the CMOS buffer-driven resistive–capacitive VLSI interconnect in subthreshold regime has been developed. This work is based on the subthreshold current model of MOS transistor. Two different cases of stimulations, viz., (i) rising input ramp and (ii) falling input ramp have been analyzed. The model so developed is used to obtain the output voltage waveform of a CMOS buffer driving large interconnect loads. The waveform shape is characterized for fast and slow ramp signals. Analytical expressions characterizing the propagation delay for rising and falling ramp signals are presented and compared with SPICE simulations. The resistive power dissipation is also quantified and verified by simulation results for 130-, 90-, and 65-nm technology nodes. The analytical driver-interconnect-load model gives sufficiently closer results to SPICE simulations.

Circuit model and characterization of CMOS buffer in subthreshold has been presented in Sect. 3.1. Analytical equations describing the waveform shape, propagation delay, and resistive power estimation for rising input ramp have been described in Sect. 3.2. Timing analysis and resistive power estimation for falling

R. Dhiman and R. Chandel, *Compact Models and Performance Investigations for Subthreshold Interconnects*, Energy Systems in Electrical Engineering,
DOI 10.1007/978-81-322-2132-6_3

ramp input have been developed in Sect. 3.3. Verification of these analytical results with SPICE is presented in Sect. 3.4. This is followed by the concluding remarks in Sect. 3.5.

3.1 Circuit Model of CMOS Buffer

It is a common practice in VLSI to drive interconnects by buffers. Buffers have been realized using CMOS inverters. The driver-interconnect-load model of the proposed analysis and equivalent circuit for the same is represented in Fig. 3.1. *Inv*1 and *Inv*2 are the driving and driven gates, respectively, C_{m} is the gate–drain coupling capacitance, V_{DD} is the supply voltage, I_{c} is the capacitive discharge current and is same as I_{n} which flows through MN, I_{m} is the gate–drain coupling current, and R is the interconnect resistance and C interconnect capacitance and includes interconnect ground capacitance and input gate capacitance of *Inv*2. In subthreshold regime, the signal frequency is quite low and of the order of kHz [12, 218]. This produces negligible inductive effect and therefore has not been considered in the present work.

The n-channel drain-to-source current of CMOS buffer in subthreshold is represented by the following expression [172]:

$$I_{\mathrm{n}} = \mu_{\mathrm{n}} C_{\mathrm{ox}} \frac{W_{\mathrm{n}}}{L_{\mathrm{n}}} (\eta_{\mathrm{n}} - 1) U_{\mathrm{th}}^2 \exp\left(\frac{V_{\mathrm{in}} - V_{\mathrm{T}}}{\eta_{\mathrm{n}} U_{\mathrm{th}}}\right) \left[1 - \exp\left(-\frac{V_{\mathrm{ds}}}{U_{\mathrm{th}}}\right)\right] \tag{3.1}$$

Here, μ_{n} is the electron mobility, C_{ox} is the gate oxide capacitance per unit area, W_{n} and L_{n} are the effective channel width and channel length, respectively, U_{th} is the thermal voltage, V_{T} is the threshold voltage, V_{in} and V_{ds} are the input voltage and drain-to-source voltage, respectively, and η_{n} is the subthreshold slope factor whose value lies between one and two. The discussion here digresses to identify two new regions in subthreshold.

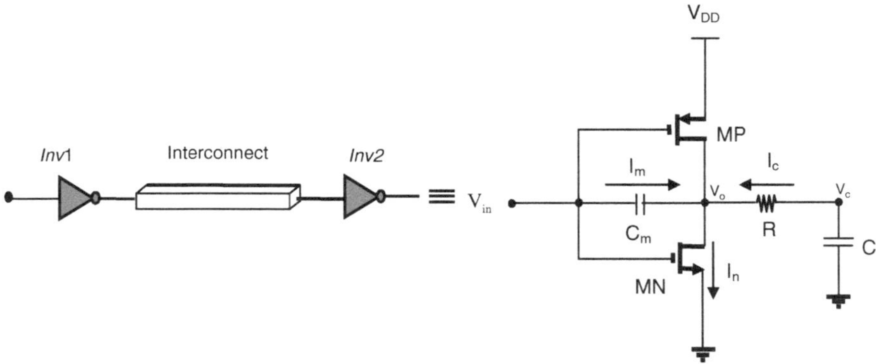

Fig. 3.1 Driver–interconnect–load model and its equivalent circuit

According to [219], for large V_{ds}, NMOS transistor is approximated by a constant current source. On the other hand, for small V_{ds}, the transistor behaves as a linear resistor. Summarizing, current in two regions can be expressed as

$$\begin{aligned} I_n &= B_n \exp\left(\frac{V_{in}-V_T}{\eta_n U_{th}}\right) && V_{ds} \geq 4U_{th}\text{: Sub-saturation region} \\ &= \gamma_n V_{ds} && V_{ds} < U_{th}\text{: Sub-linear region} \end{aligned} \tag{3.2}$$

In Eq. (3.2), B_n and γ_n are given as

$$B_n = \mu_n C_{ox} \frac{W_n}{L_n} (\eta_n - 1) U_{th}^2 \tag{3.3}$$

$$\gamma_n = \mu_n C_{ox} \frac{W_n}{L_n} (\eta_n - 1) U_{th} \tag{3.4}$$

B_n is the drain-to-source source current when $V_{in} = V_T$, and γ_n is the output conductance of MN in the sub-linear region. B_n and γ_n have the units of current and transconductance, respectively.

3.2 Analysis of Buffer-Driven Interconnect for Rising Ramp Input

Application of KCL in Fig. 3.1 gives,

$$+I_m + I_c - I_n = 0 \tag{3.5}$$

For a single transistor primitive of a CMOS inverter and a rising input ramp, (3.5) may be written as

$$+C_m \left(\frac{dV_{in}}{dt} - \frac{dV_o}{dt} \right) - C \frac{dV_c}{dt} - I_n = 0 \tag{3.6}$$

where V_o and V_c are the drain-to-source voltage and capacitor voltage, respectively.
Application of KVL in Fig. 3.1 results in

$$V_o = V_c + RC \frac{dV_c}{dt} \tag{3.7}$$

Substituting (3.7) in (3.6) gives

$$\frac{dV_c}{dt} + C_b \frac{d^2 V_c}{dt^2} - C_3 \frac{dV_{in}}{dt} + \frac{I_n}{C_a} = 0 \tag{3.8}$$

where

$$C_{\mathrm{a}} = C + C_{\mathrm{m}}, \quad C_{\mathrm{b}} = \frac{RCC_{\mathrm{m}}}{C_{\mathrm{a}}}, \quad C_3 = \frac{C_{\mathrm{m}}}{C_{\mathrm{a}}} \tag{3.9}$$

For rising ramp input, two main cases are considered: fast ramp and slow ramp. Depending on the status of the active device, the input ramp is categorized to fast or slow. If the active device continues to operate in the sub-saturation region even when the input has completed its transition, the ramp is said to be fast. On the other hand, if the device enters into the sub-linear region before the completion of input transition, the input ramp signal is called slow. For a fixed-size MOS, the region of operation also depends on the load characteristics. If the interconnect is long; that is, the parasitic impedance parameters are large, the discharging through the n-channel transistor (NMOS) will be slow, thereby compelling the transistor to continue its operation in sub-saturation region. The input ramp in this case is therefore categorized as fast ramp. Contrary to this condition, keeping all other parameters same, if interconnect length is short, NMOS discharges the load capacitance quick enough to enter sub-linear region of operation before the ramp completes the transition. Under this condition, the input ramp is treated as slow ramp.

In the subsequent analysis, the output voltage is determined based on the fast and slow ramp input signals. Furthermore, for the purpose of analysis, it is assumed that threshold voltages of the constituent transistors are equal.

3.2.1 Fast Ramp

The input is assumed to be a rising ramp signal with rise time τ_{r} defined as

$$\begin{aligned} V_{\mathrm{in}} &= V_{\mathrm{DD}} \frac{t}{\tau_{\mathrm{r}}} \quad \text{for } 0 \leq t \leq \tau_{\mathrm{r}}, \\ &= V_{\mathrm{DD}} \quad \text{for } t > \tau_{\mathrm{r}} \end{aligned} \tag{3.10}$$

The time dependence of V_{o} is obtained from (3.10) over different regions of input transition, as presented below.

Region-1 $(0 \leq t \leq \tau_{\mathrm{r}})$: It is assumed that input transition is fast enough to keep the NMOS transistor (MN) in sub-saturation over this period of time. Applying the boundary condition that at $t = 0$, $V_{\mathrm{o}} = V_{\mathrm{DD}}$ and substituting for I_{n}, (3.8) is solved to obtain

$$\begin{aligned} V_{\mathrm{o}} = {} & V_{\mathrm{DD}}\left[1 - \mathrm{e}^{-\frac{t}{C_{\mathrm{b}}}}\right] - \frac{B_{\mathrm{n}}}{C_{\mathrm{a}}}\frac{\tau_{\mathrm{r}}\eta_{\mathrm{n}}U_{\mathrm{th}}}{V_{\mathrm{DD}}}\left[\mathrm{e}^{\frac{V_{\mathrm{DD}}\frac{t}{\tau_{\mathrm{r}}} - V_{\mathrm{DD}}}{\eta_{\mathrm{n}}U_{\mathrm{th}}}} - \mathrm{e}^{-\frac{V_{\mathrm{DD}}}{\eta_{\mathrm{n}}U_{\mathrm{th}}}}\right] - RB_{\mathrm{n}}\left[\mathrm{e}^{\frac{V_{\mathrm{DD}}\frac{t}{\tau_{\mathrm{r}}} - V_{\mathrm{DD}}}{\eta_{\mathrm{n}}U_{\mathrm{th}}}} - \mathrm{e}^{-\frac{V_{\mathrm{DD}}}{\eta_{\mathrm{n}}U_{\mathrm{th}}}}\right] \\ & + \frac{C_3 V_{\mathrm{DD}}}{\tau_{\mathrm{r}}}\left[t - C_{\mathrm{b}}\left(1 - \mathrm{e}^{-\frac{t}{C_{\mathrm{b}}}}\right)\right] + V_{\mathrm{DD}}\mathrm{e}^{-\frac{t}{C_{\mathrm{b}}}} \end{aligned} \tag{3.11}$$

For submicron and deep submicron devices, C_m is very small, and for a long interconnect, line-to-ground capacitance is very large. Thus, from (3.8), it can be assumed that $C_a \cong C$, $C_b = C_3 \cong 0$. This reduces (3.11) to

$$V_o = V_{DD} - \frac{B_n}{C}\frac{\tau_r \eta_n U_{th}}{V_{DD}}\left[e^{\frac{V_{DD}\frac{t}{\tau_r} - V_{DD}}{\eta_n U_{th}}} - e^{-\frac{V_{DD}}{\eta_n U_{th}}}\right] - RB_n\left[e^{\frac{V_{DD}\frac{t}{\tau_r} - V_{DD}}{\eta_n U_{th}}} - e^{-\frac{V_{DD}}{\eta_n U_{th}}}\right] \quad (3.12)$$

At $t = \tau_r$, the output voltage obtained is

$$V_o(\tau_r) = V_{DD} - \frac{B_n}{C}\frac{\tau_r \eta_n U_{th}}{V_{DD}}\left[1 - e^{-\frac{V_{DD}}{\eta_n U_{th}}}\right] - RB_n\left[1 - e^{-\frac{V_{DD}}{\eta_n U_{th}}}\right] \quad (3.13)$$

Region-2 ($\tau_r \le t \le \tau_{nsat}$): In region-2, the transition of fast input signal is completed and the input voltage is fixed at V_{DD}. MN remains in the sub-saturation region of its characteristics. The drain-to-source current of MN is $I_n = B_n$. The time duration when MN leaves the sub-saturation region of its characteristics is denoted by τ_{nsat}. The output voltage is obtained based on the condition at $t = \tau_r$ and is given as

$$V_o = V_{DD} - \frac{B_n}{C}\frac{\tau_r \eta_n U_{th}}{V_{DD}}\left[1 - e^{-\frac{V_{DD}}{\eta_n U_{th}}}\right] - RB_n\left[1 - e^{-\frac{V_{DD}}{\eta_n U_{th}}}\right] - \frac{B_n}{C}(t - \tau_r) \quad (3.14)$$

Region-3 ($t > \tau_{nsat}$): In this region, the output voltage falls below $4U_{th}$ and MN enters into the sub-linear region. Substituting for I_n, differential equation (3.8) is solved to yield the output voltage expression in region-3 as

$$V_o = V_o(\tau_{nsat}) e^{-\frac{\gamma_n}{C(1+\gamma_n R)}(t - \tau_{nsat})} \quad (3.15)$$

Equation (3.15) is in accordance with that reported in [220], for an RC interconnect load. Thus, the shape of the output voltage waveform is dependent upon interconnect load and device parameters as can be seen from Eqs. (3.12)–(3.15).

3.2.2 Determination of τ_{nsat}

At $t = \tau_{nsat}$, MN is just at the end of its sub-saturation region and from (3.14)

$$V_o(\tau_{nsat}) = V_{DD} - \frac{B_n}{C}\frac{\tau_r \eta_n U_{th}}{V_{DD}}\left[1 - e^{-\frac{V_{DD}}{\eta_n U_{th}}}\right] - RB_n\left[1 - e^{-\frac{V_{DD}}{\eta_n U_{th}}}\right] - \frac{B_n}{C}(\tau_{nsat} - \tau_r) \quad (3.16)$$

According to [224], at $t = \tau_{nsat}$,

$$V_o = 4U_{th} \tag{3.17}$$

Substituting (3.17) into (3.16) to yield

$$\tau_{nsat} = \tau_r + \frac{C}{B_n}(V_{DD} - 4U_{th}) - \left(1 - e^{-\frac{V_{DD}}{\eta_n U_{th}}}\right)\left(RC + \frac{\tau_r \eta_n U_{th}}{V_{DD}}\right). \tag{3.18}$$

3.2.3 High-to-Low Propagation Delay of a Fast Ramp Signal

The high-to-low propagation delay $(t_{P_{HL}})$ is defined as the time required by the output voltage to reach 50 % of its initial value. It is approximated using (3.14) and is given as

$$t_{P_{HL}} = \tau_r + 0.5V_{DD}\frac{C}{B_n} - \left(1 - e^{-\frac{V_{DD}}{\eta_n U_{th}}}\right)\left(RC + \frac{\tau_r \eta_n U_{th}}{V_{DD}}\right) \tag{3.19}$$

Note that there are three terms in the delay expression of (3.19). In the first term, delay is proportional to the input rise time and hence is waveform shape dependent. The other terms in the expression show the dependence of delay on power supply, device parameters, and load conditions. On the similar lines, ninety percent high-to-low propagation delay $(t_{0.1})$ is given by

$$t_{0.1} = \tau_r + 0.9V_{DD}\frac{C}{B_n} - \left(1 - e^{-\frac{V_{DD}}{\eta_n U_{th}}}\right)\left(RC + \frac{\tau_r \eta_n U_{th}}{V_{DD}}\right) \tag{3.20}$$

Thus, we note that propagation delay of a submicrometer CMOS buffer driving an interconnect load is a function of various parameters. It may be expressed as a function of the various parameters in the following way:

$$\text{Delay (High-to-Low)} = f(\tau_r, V_{DD}, R, C, B_n). \tag{3.21}$$

3.2.4 High-to-Low Propagation Delay of a Slow Ramp Signal

The analysis described earlier is presented for a fast ramp input signal. In the following section, the analysis is based on the assumption of a slow ramp signal; that is,

MN enters into the sub-linear region before the completion of input transition. According to [221], criterion for fast ramp input signal is

$$f(\tau_r) = \tau_{nsat} - \tau_r \geq 0 \tag{3.22}$$

$$= \frac{C}{B_n}(V_{DD} - 4U_{th}) - \left(1 - e^{-\frac{V_{DD}}{\eta_n U_{th}}}\right)\left(RC + \frac{\tau_r \eta_n U_{th}}{V_{DD}}\right) \geq 0 \tag{3.23}$$

Otherwise, the input signal is a slow ramp. The output voltage for a slow ramp signal in the region $0 \leq t \leq \tau_{nsat}$ is same as (3.12). MN leaves the sub-saturation region at $t = \tau_{nsat}$ and in this case is approximated based upon

$$V_{DD} - \frac{B_n}{C}\frac{\tau_r \eta_n U_{th}}{V_{DD}}\left[e^{\frac{V_{DD}\frac{\tau_{nsat}}{\tau_r} - V_{DD}}{\eta_n U_{th}}} - e^{-\frac{V_{DD}}{\eta_n U_{th}}}\right] - RB_n\left[e^{\frac{V_{DD}\frac{\tau_{nsat}}{\tau_r} - V_{DD}}{\eta_n U_{th}}} - e^{-\frac{V_{DD}}{\eta_n U_{th}}}\right] = 4U_{th} \tag{3.24}$$

The 50 % high-to-low propagation delay for slow ramp $(t_{0.5})$ is approximated using (3.12) as

$$\frac{B_n}{C}\frac{\tau_r \eta_n U_{th}}{V_{DD}}\left[e^{\frac{V_{DD}\frac{t_{0.5}}{\tau_r} - V_{DD}}{\eta_n U_{th}}} - e^{-\frac{V_{DD}}{\eta_n U_{th}}}\right] - RB_n\left[e^{\frac{V_{DD}\frac{t_{0.5}}{\tau_r} - V_{DD}}{\eta_n U_{th}}} - e^{-\frac{V_{DD}}{\eta_n U_{th}}}\right] = 0.5V_{DD} \tag{3.25}$$

The solutions of τ_{nsat} and $t_{0.5}$ have been obtained using Newton–Raphson solver. Thus, the high-to-low propagation delays for both fast and slow ramp signals have been derived analytically in (3.19) and (3.25).

3.2.5 Power Estimation

Power dissipation is also an important concern in VLSI circuits. Millions of active devices and interconnections make managing delay and power dissipation a challenging task in VLSI design. The main sources of power dissipation in VLSI circuits are static power and dynamic power. In addition, interconnect resistance also dissipates power. The power dissipation across interconnect resistance is termed as resistive power dissipation (P_R). Resistive power dissipation can be quantified as

$$P_R = f\int_0^t I_n^2 R\,dt \tag{3.26}$$

f is the frequency of operation.

In region-1, power dissipated by the interconnect resistance is computed as

$$P_{\mathrm{R-1}} = fRB_n^2 \mathrm{e}^{-\frac{2V_{\mathrm{DD}}}{\eta_{\mathrm{n}} U_{\mathrm{th}}}} \frac{\tau_{\mathrm{r}} \eta_{\mathrm{n}} U_{\mathrm{th}}}{2V_{\mathrm{DD}}} \left(\mathrm{e}^{\frac{2V_{\mathrm{DD}}}{\eta_{\mathrm{n}} U_{\mathrm{th}}}} - 1 \right) \tag{3.27}$$

In region-2, MN operates with the constant discharge current which is equal to B_{n}. The resistive power dissipation is given by

$$P_{\mathrm{R-2}} = fR \int_{\tau_{\mathrm{r}}}^{\tau_{\mathrm{nsat}}} B_{\mathrm{n}}^2 \mathrm{d}t = fRB_{\mathrm{n}}^2 (\tau_{\mathrm{nsat}} - \tau_{\mathrm{r}}) \tag{3.28}$$

In region-3, the resistive power dissipation can be calculated by solving (3.26) within the time limits as

$$P_{\mathrm{R-3}} = fR\gamma_{\mathrm{n}}^2 \int_{\tau_{\mathrm{nsat}}}^{t_{0.1}} V_{\mathrm{o}}^2 \mathrm{d}t = \frac{fRC\gamma_{\mathrm{n}} V_{\mathrm{o}}^2(\tau_{\mathrm{nsat}})}{2(1+\gamma_{\mathrm{n}} R)} \left[1 - \mathrm{e}^{-\frac{2\gamma_{\mathrm{n}}}{C(1+\gamma_{\mathrm{n}} R)}(t_{0.1} - \tau_{\mathrm{nsat}})} \right] \tag{3.29}$$

Thus, the power dissipated by interconnect resistance during high-to-low output transition ($P_{\mathrm{R_{HL}}}$) is given as

$$P_{\mathrm{R_{HL}}} = P_{\mathrm{R-1}} + P_{\mathrm{R-2}} + P_{\mathrm{R-3}} \tag{3.30}$$

The root mean square value of resistive current in each of the three regions can be expressed as

$$I_{\mathrm{n-1}} = \sqrt{fB_{\mathrm{n}}^2 \mathrm{e}^{-\frac{2V_{\mathrm{DD}}}{\eta_{\mathrm{n}} U_{\mathrm{th}}}} \frac{\tau_{\mathrm{r}} \eta_{\mathrm{n}} U_{\mathrm{th}}}{2V_{\mathrm{DD}}} \left(\mathrm{e}^{\frac{2V_{\mathrm{DD}}}{\eta_{\mathrm{n}} U_{\mathrm{th}}}} - 1 \right)} \tag{3.31}$$

$$I_{\mathrm{n-2}} = \sqrt{fB_{\mathrm{n}}^2 (\tau_{\mathrm{nsat}} - \tau_{\mathrm{r}})} \tag{3.32}$$

$$I_{\mathrm{n-3}} = \sqrt{\frac{fC\gamma_{\mathrm{n}} V_{\mathrm{o}}^2(\tau_{\mathrm{nsat}})}{2(1+\gamma_{\mathrm{n}} R)} \left[1 - \mathrm{e}^{-\frac{2\gamma_{\mathrm{n}}}{C(1+\gamma_{\mathrm{n}} R)}(t_{0.1} - \tau_{\mathrm{nsat}})} \right]} \tag{3.33}$$

The root mean square resistive current $(I_{\mathrm{n,\,rms}})$ that flows during high-to-low output transition is therefore

$$I_{\mathrm{n,rms}} = \sqrt{I_{\mathrm{n-1}}^2 + I_{\mathrm{n-2}}^2 + I_{\mathrm{n-3}}^2} \tag{3.34}$$

3.3 Analysis for Falling Ramp Input

A similar analysis as in Sect. 3.2 gives the output voltage of a buffer-driven RC interconnect for a falling ramp input. The time dependence of V_o in the various regions of input transition is obtained by solving (3.8) with PMOS transistor (MP) as the active device.

3.3.1 Fast Ramp

The falling input ramp signal is characterized by

$$\begin{aligned} V_{in} &= V_{DD}\left(1-\frac{t}{\tau_r}\right) \quad && \text{for } 0 \le t \le \tau_r, \\ &= 0 && \text{for } t > \tau_r \end{aligned} \tag{3.35}$$

Region-1 ($0 \le t \le \tau_r$): Over this interval of time, MP is in the sub-saturation region. Applying the boundary condition; that is, at $t = 0$, $V_o = 0$, the output voltage is given by,

$$V_o = \frac{B_p}{C}\frac{\tau_r \eta_p U_{th}}{V_{DD}}\left[e^{\frac{V_{DD}\frac{t}{\tau_r}-V_{DD}}{\eta_p U_{th}}} - e^{-\frac{V_{DD}}{\eta_p U_{th}}}\right] + RB_p\left[e^{\frac{V_{DD}\frac{t}{\tau_r}-V_{DD}}{\eta_p U_{th}}} - e^{-\frac{V_{DD}}{\eta_p U_{th}}}\right] \tag{3.36}$$

Region-2 ($\tau_r \le t \le \tau_{psat}$): MP is in the sub-saturation region of its characteristics. The source-to-drain current of MP is $I_p = B_p$. As $V_{in} = 0$, (3.8) is solved to obtain

$$V_o = V_o(\tau_r) + \frac{B_p}{C}(t-\tau_r) \tag{3.37}$$

The output voltage obtained based on the condition at $t = \tau_r$ is given by

$$V_o(\tau_r) = \frac{B_p}{C}\frac{\tau_r \eta_p U_{th}}{V_{DD}}\left[1 - e^{-\frac{V_{DD}}{\eta_p U_{th}}}\right] + RB_p\left[1 - e^{-\frac{V_{DD}}{\eta_p U_{th}}}\right] \tag{3.38}$$

Region-3 ($t > \tau_{psat}$): MP leaves the sub-saturation at $t = \tau_{psat}$ and enters into the sub-linear region. The output voltage is governed by

$$V_o = V_{DD} + \left(V_o(\tau_{psat}) - V_{DD}\right)e^{-\frac{\gamma_p}{C(1+\gamma_p R)}\left(t-\tau_{psat}\right)}. \tag{3.39}$$

3.3.2 Determination of τ_{psat}

At $t = \tau_{\mathrm{psat}}$, sub-saturation drain current is equal to sub-linear drain current. Thus, from (3.37), we have

$$V_{\mathrm{o}}(\tau_{\mathrm{r}}) + \frac{B_{\mathrm{p}}}{C}(\tau_{\mathrm{psat}} - \tau_{\mathrm{r}}) = V_{\mathrm{DD}} - 4U_{\mathrm{th}} \tag{3.40}$$

which is solved to yield

$$\tau_{\mathrm{psat}} = \tau_{\mathrm{r}} + \frac{C}{B_{\mathrm{p}}}(V_{\mathrm{DD}} - 4U_{\mathrm{th}}) - \left(1 - \mathrm{e}^{-\frac{V_{\mathrm{DD}}}{\eta_{\mathrm{p}}U_{\mathrm{th}}}}\right)\left(RC + \frac{\tau_{\mathrm{r}}\eta_{\mathrm{p}}U_{\mathrm{th}}}{V_{\mathrm{DD}}}\right). \tag{3.41}$$

3.3.3 Low-to-High Propagation Delay of a Fast Ramp Signal

The low-to-high propagation delay $(t_{\mathrm{P_{LH}}})$ is approximated using (3.37) as

$$V_{\mathrm{o}}(\tau_{\mathrm{r}}) + \frac{B_{\mathrm{p}}}{C}(t_{\mathrm{P_{LH}}} - \tau_{\mathrm{r}}) = 0.5V_{\mathrm{DD}} \tag{3.42}$$

Equation (3.42) is solved to yield

$$t_{\mathrm{P_{LH}}} = \tau_{\mathrm{r}} + 0.5V_{\mathrm{DD}}\frac{C}{B_{\mathrm{p}}} - \frac{\tau_{\mathrm{r}}\eta_{\mathrm{p}}U_{\mathrm{th}}}{V_{\mathrm{DD}}}\left(1 - \mathrm{e}^{-\frac{V_{\mathrm{DD}}}{\eta_{\mathrm{p}}U_{\mathrm{th}}}}\right) - RC\left(1 - \mathrm{e}^{-\frac{V_{\mathrm{DD}}}{\eta_{\mathrm{p}}U_{\mathrm{th}}}}\right). \tag{3.43}$$

3.3.4 Low-to-High Propagation Delay of a Slow Ramp Signal

The output voltage for slow ramp signal in the time domain $0 \le t \le \tau_{\mathrm{psat}}$ is same as (3.36). In this case, however, τ_{psat} is calculated based upon

$$\frac{B_{\mathrm{p}}}{C}\frac{\tau_{\mathrm{r}}\eta_{\mathrm{p}}U_{\mathrm{th}}}{V_{\mathrm{DD}}}\left[\mathrm{e}^{\frac{V_{\mathrm{DD}}\frac{\tau_{\mathrm{psat}}}{\tau_{\mathrm{r}}} - V_{\mathrm{DD}}}{\eta_{\mathrm{p}}U_{\mathrm{th}}}} - \mathrm{e}^{-\frac{V_{\mathrm{DD}}}{\eta_{\mathrm{p}}U_{\mathrm{th}}}}\right] + RB_{\mathrm{p}}\left[\mathrm{e}^{\frac{V_{\mathrm{DD}}\frac{\tau_{\mathrm{psat}}}{\tau_{\mathrm{r}}} - V_{\mathrm{DD}}}{\eta_{\mathrm{p}}U_{\mathrm{th}}}} - \mathrm{e}^{-\frac{V_{\mathrm{DD}}}{\eta_{\mathrm{p}}U_{\mathrm{th}}}}\right] = V_{DD} - 4U_{th} \tag{3.44}$$

The low-to-high propagation delay for slow ramp signal $(t_{0.5})$ is also approximated using (3.36) as

$$\frac{B_{\mathrm{p}}}{C}\frac{\tau_{\mathrm{r}}\eta_{\mathrm{p}}U_{\mathrm{th}}}{V_{\mathrm{DD}}}\left[\mathrm{e}^{\frac{V_{\mathrm{DD}}\frac{t_{0.5}}{\tau_{\mathrm{r}}} - V_{\mathrm{DD}}}{\eta_{\mathrm{p}}U_{\mathrm{th}}}} - \mathrm{e}^{-\frac{V_{\mathrm{DD}}}{\eta_{\mathrm{p}}U_{\mathrm{th}}}}\right] + RB_{\mathrm{p}}\left[\mathrm{e}^{\frac{V_{\mathrm{DD}}\frac{t_{0.5}}{\tau_{\mathrm{r}}} - V_{\mathrm{DD}}}{\eta_{\mathrm{p}}U_{\mathrm{th}}}} - \mathrm{e}^{-\frac{V_{\mathrm{DD}}}{\eta_{\mathrm{p}}U_{\mathrm{th}}}}\right] = 0.5V_{\mathrm{DD}} \tag{3.45}$$

Newton–Raphson numeric solver has been used in (3.44) and (3.45) to obtain τ_{psat} and $t_{0.5}$.

3.3.5 Resistive Power Dissipation

In region-1, the power dissipated by the interconnect resistance is given by

$$P_{R-1} = fRB_p^2 \int_0^{\tau_r} \left(e^{\frac{V_{DD}\frac{t}{\tau_r} - V_{DD}}{\eta_p U_{th}}} \right)^2 dt = fRB_p^2 e^{-\frac{2V_{DD}}{\eta_p U_{th}}} \frac{\tau_r \eta_p U_{th}}{2V_{DD}} \left(e^{\frac{2V_{DD}}{\eta_p U_{th}}} - 1 \right) \quad (3.46)$$

In region-2, MP operates in sub-saturation with constant discharge current equal to B_p. The resistive power dissipation is given by

$$P_{R-2} = f\mathrm{R} \int_{\tau_r}^{\tau_{psat}} B_p^2 dt = fRB_p^2 (\tau_{psat} - \tau_r) \quad (3.47)$$

In region-3, the resistive power dissipation is calculated as

$$P_{R-3} = fRC\gamma_p \frac{\{V_o(\tau_{psat}) - V_{DD}\}^2}{2(1 + \gamma_p \mathrm{R})} \left[1 - e^{-\frac{2\gamma_p}{C(1+\gamma_p \mathrm{R})}(t_{0.9} - \tau_{psat})} \right] \quad (3.48)$$

The root mean square resistive current in each of the three regions can therefore be expressed as

$$I_{p-1} = \sqrt{fB_p^2 e^{-\frac{2V_{DD}}{\eta_p U_{th}}} \frac{\tau_r \eta_p U_{th}}{2V_{DD}} \left(e^{\frac{2V_{DD}}{\eta_p U_{th}}} - 1 \right)} \quad (3.49)$$

$$I_{p-2} = \sqrt{fB_p^2 (\tau_{psat} - \tau_r)} \quad (3.50)$$

$$I_{p-3} = \sqrt{fC\gamma_p \frac{\{V_o(\tau_{psat}) - V_{DD}\}^2}{2(1 + \gamma_p R)} \left[1 - e^{-\frac{2\gamma_p}{C(1+\gamma_p \mathrm{R})}(t_{0.9} - \tau_{psat})} \right]} \quad (3.51)$$

The root mean square resistive current $(I_{p,rms})$ that flows during low-to-high output transition is therefore

$$I_{p,rms} = \sqrt{I_{p-1}^2 + I_{p-2}^2 + I_{p-3}^2}. \quad (3.52)$$

3.4 Comparison with Simulation Results

Output voltage waveforms are obtained by the present analytical approach and SPICE simulation results. The values of supply voltage are 0.30, 0.34, and 0.36 V, respectively, in 130-, 90-, and 65-nm technology nodes. The device parameters and impedance parasitics of Cu interconnect are extracted using BSIM level 54 MOS parameters and interconnect geometries suggested in [222] for the technology nodes considered in the present work. The time dependence of output voltage for fast ramp input to CMOS buffer-driven RC interconnect is shown in Fig. 3.2a–d. Interconnect length of 5 mm is taken. The input ramp has a rise time of 0.1 μs. The frequency of

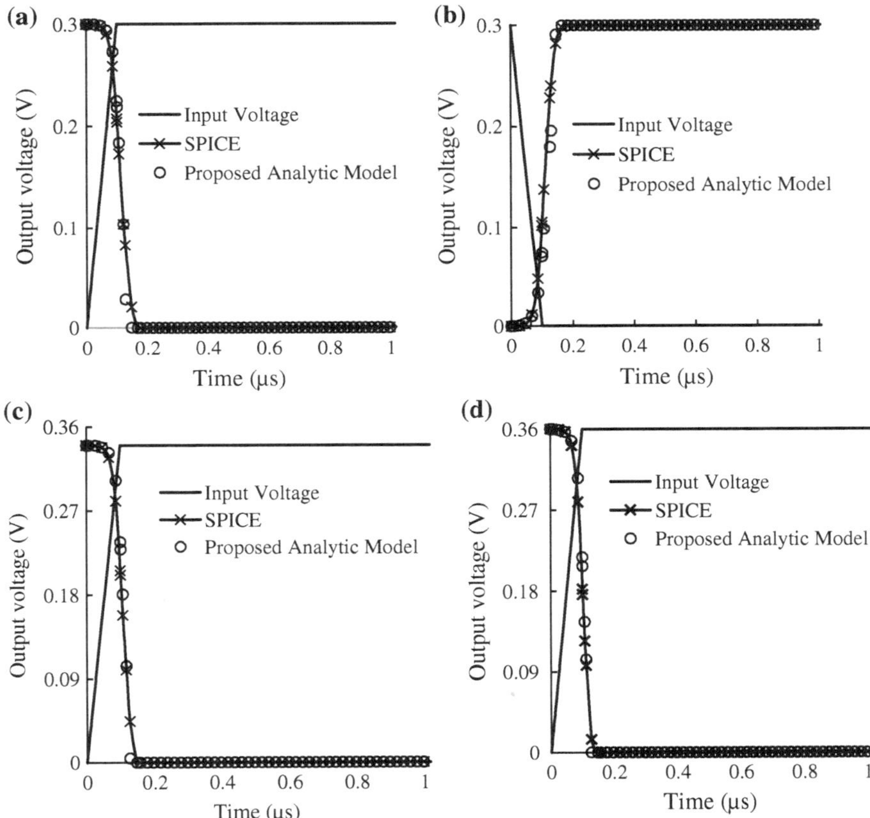

Fig. 3.2 Analytically obtained time dependence of output voltage and SPICE transient simulation results under fast ramp for three technology nodes **a**, **b** 130 nm, **c** 90 nm, **d** 65 nm. **a**, **b** W_n = 195 nm, $W_p = 2.5W_n$, R = 160.82 Ω, C = 187.47 fF. **c** W_n = 135 nm, $W_p = 2.5W_n$, R = 209.57 Ω, C = 195.88 fF. **d** W_n = 97.5 nm, $W_p = 2.5W_n$, R = 208.93 Ω, C = 139.02 fF

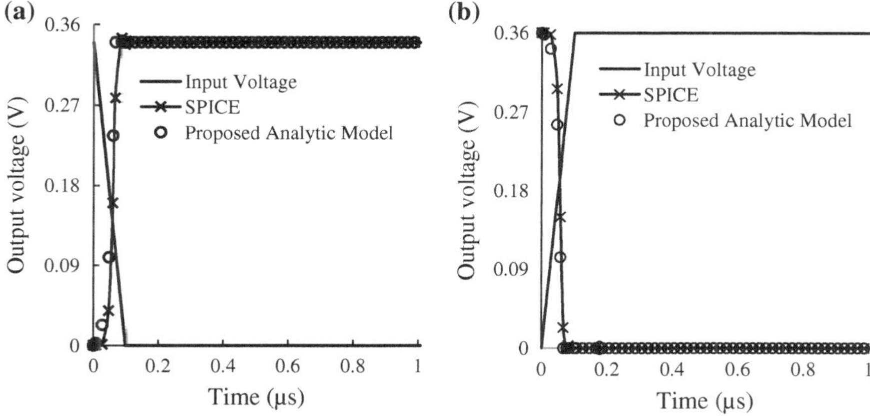

Fig. 3.3 Analytical and SPICE output voltages under slow ramp with time for **a** 90-nm and **b** 65-nm technology nodes. **a** $W_n = 8$ μm, $W_p = 2.5W_n$, $R = 209.57$ Ω, $C = 195.88$ fF. **b** $W_n = 10$ μm, $W_p = 2.5W_n$, $R = 208.93$ Ω, $C = 139.02$ fF

the input ramp is 100 kHz. From Fig. 3.2a–d, it is seen that there is a good agreement between the analytically obtained output and SPICE simulations.

SPICE and analytical voltage waveforms for slow ramp signal in 90 and 65 nm technology nodes have been shown in Fig. 3.3a–b, respectively. Note that the output voltage based on the analytical expressions is quite close to SPICE simulation for each condition. The analytical expressions therefore can be used to predict the waveform shape of the output voltage signal.

The analytic variation of output voltage with time for different NMOS widths in 65 nm technology is shown in Fig. 3.4. PMOS width is 2.5 times of W_n. It is seen

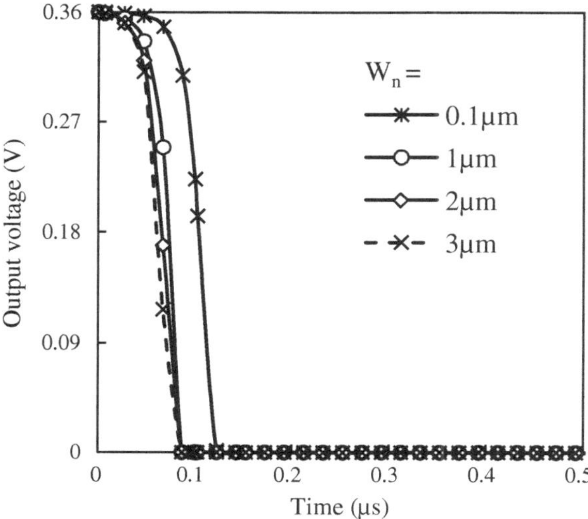

Fig. 3.4 Variation of analytical output voltage with time for different W_n for 65-nm technology node

Table 3.1 Computational error involved in timing and high-to-low propagation delay

Technology node (nm)	Ramp type	W_n (μm)	Timing (μs)			Delay (μs)		
			SPICE	Proposed model	$\varepsilon_{50\%}$ (%)	SPICE	Proposed model	$\varepsilon_{50\%}$ (%)
130	Fast	0.195	0.123	0.121	1.71	0.113	0.114	0.35
	Slow	12	0.066	0.062	5.48	0.063	0.059	6.76
90	Fast	0.135	0.115	0.117	1.57	0.106	0.109	2.26
	Slow	8	0.064	0.063	1.83	0.061	0.059	3.61
65	Fast	0.0975	0.109	0.111	1.93	0.101	0.104	2.67
	Slow	2	0.070	0.073	4.29	0.065	0.070	7.69

from Fig. 3.4 that the percentage variation in output voltage is very nominal beyond 2 μm width. This is because the advantage in delay minimization is limited with larger buffer size. In the present case, 2 μm is the optimum channel width of the NMOS transistor. Analytical result is in good agreement with the known fact that beyond a threshold channel width, there is no further improvement in delay. Thus, optimum-sized buffer in subthreshold regime is preferred to meet the ultra-low-power requirement and delay design constraints.

Next, 50 % propagation delay is analytically determined for rising and falling ramps. Both fast and slow ramp inputs have been considered in the present analysis which categorically depends upon the width of the active device. These results have been provided in Tables 3.1 and 3.2 along with the percentage error ($\varepsilon_{50\%}$) with respect to (wrt) SPICE simulation results. The time when the active device makes transition from sub-saturation to sub-linear region is also calculated through the proposed models and is compared with the SPICE simulations. The same is specified (as Timing) in Tables 3.1 and 3.2.

It may be seen from Table 3.1 that the proposed analytical model yields maximum percentage errors as 5.48 and 7.69 % in the timing and high-to-low propagation delay, respectively. The average percentage error involved in the same are 2.80 and 3.89 %, respectively. The errors involved in timing and low-to-high propagation delay predicted by the proposed models are provided in Table 3.2. For test cases

Table 3.2 Computational error involved in timing and low-to-high propagation delay

Technology node (nm)	Ramp type	W_n (μm)	Timing (μs)			Delay (μs)		
			SPICE	Proposed model	$\varepsilon_{50\%}$ (%)	SPICE	Proposed model	$\varepsilon_{50\%}$ (%)
130	Fast	0.195	0.119	0.124	04.20	0.110	0.120	08.64
	Slow	12	0.064	0.062	02.36	0.061	0.058	04.91
90	Fast	0.135	0.109	0.128	17.16	0.101	0.115	14.06
	Slow	8	0.064	0.060	06.72	0.060	0.055	08.62
65	Fast	0.0975	0.111	0.118	06.31	0.101	0.112	10.89
	Slow	2	0.072	0.068	05.56	0.066	0.063	04.27

under consideration, the maximum percentage errors are 17.16 and 14.06 %, while average percentage errors are 7.05 and 8.57 % in timing and low-to-high propagation delay, respectively.

The variation of propagation delay and conductance with buffer width (W_n) for the three technology nodes is shown in Fig. 3.5a–c. Interconnect length of 1 mm has been considered. The channel width of PMOS is 2.5 times that of NMOS width.

It can be observed that as the size of NMOS transistor is increased, conductance increases, while propagation delay decreases to a minimum and then levels off. For larger channel widths and hence the aspect ratio, γ_n is higher as it is directly proportional to the aspect ratio; this can be seen in Eq. (3.4). This increases the current driving capability of the buffer, and the load discharge is faster, leading to lower delays.

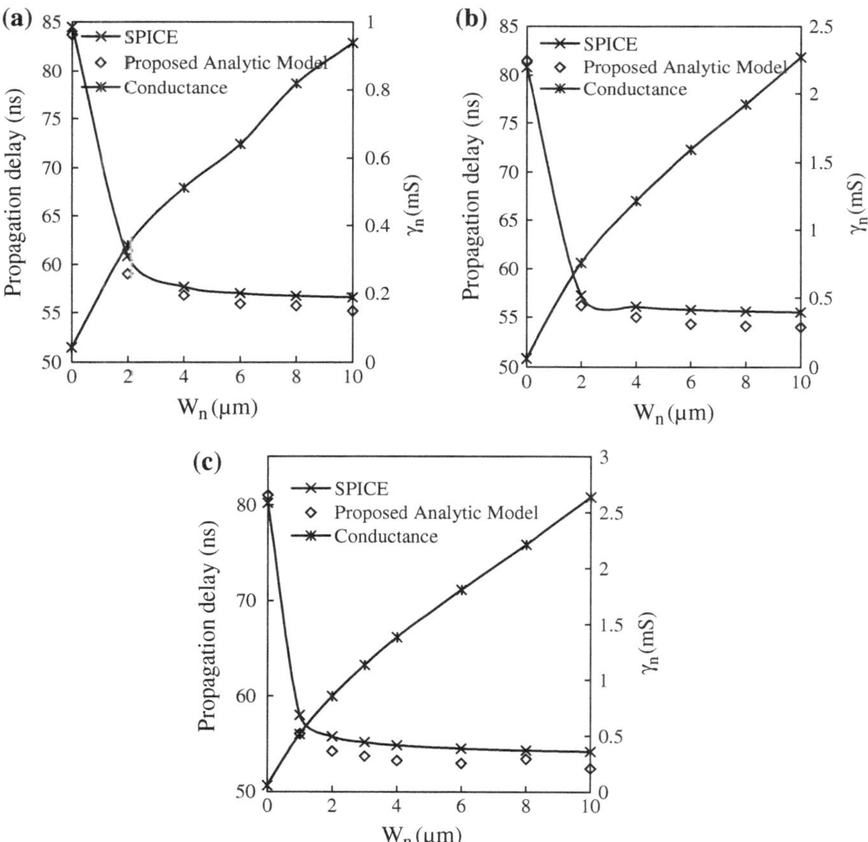

Fig. 3.5 Variation of drain conductance, SPICE, and proposed analytical propagation delay with NMOS width for **a** 130-nm, **b** 90-nm, and **c** 65-nm technology nodes

Table 3.3 Propagation delay (ns) with interconnect length for different technology nodes

L_i (μm)	130-nm technology			90-nm technology			65-nm technology		
	SPICE	Proposed model	$\varepsilon_{50\%}$ (%)	SPICE	Proposed model	$\varepsilon_{50\%}$ (%)	SPICE	Proposed model	$\varepsilon_{50\%}$ (%)
600	77.19	81.43	5.49	75.25	78.94	4.90	74.73	72.40	3.12
700	78.97	83.45	5.67	76.57	80.94	5.71	76.06	74.45	2.12
800	80.80	85.16	5.40	77.97	82.67	6.03	77.44	75.42	2.61
900	82.64	86.69	4.90	79.40	84.20	6.05	78.82	76.66	2.74
1,000	84.47	83.7	1.61	80.84	85.57	5.85	80.20	78.92	1.60

The variation of propagation delay with the length of interconnect (L_i) is given in Table 3.3. L_i is varied from 600 to 1,000 μm [223]. It is seen from Table 3.3 that maximum percentage error in propagation delay predicted by the proposed model wrt SPICE for 130-nm technology is 5.67 and 6.05 % in 90-nm technology. The maximum percentage error for 65-nm technology is 3.12 %, which is lowest among the three technology nodes. It is therefore encouraging to note that the proposed delay model is more accurate for short-channel devices. The reason is that the MOSFET model utilized is more appropriate for deep submicron technologies. As a result, the proposed model gives more accurate results in the lowest of the three technologies. The average percentage errors in the estimation of the same are 4.61, 5.71, and 2.44 % for 130-, 90-, and 65-nm technology nodes, respectively. It is also observed that the propagation delay is lesser for smaller interconnect lengths.

An important result of the analysis is that increase in delay with L_i is technology dependent. It is seen from Table 3.3 that increase in delay for 130-nm technology is 9.43 % as interconnect length increases from 600 to 1,000 μm. In case of 90-nm technology, this increase is 7.43 and is 7.31 % for 65-nm technology, respectively. Thus, as the technology scales down, propagation delay of MOS device in subthreshold improves.

Low-power VLSI designs and biomedical applications need to contain power dissipation of the circuits quite stringently. The variation of power dissipation in subthreshold (st) and super-threshold (ST) with W_n for three technology nodes is shown in Fig. 3.6a–c as obtained from SPICE simulations. Variation in W_n also reflects the corresponding variation in W_p.

It is observed that operating CMOS-driven VLSI interconnects in subthreshold regime leads to savings in power dissipation by over an order. For instance, in case of 65-nm technology node for $W_n = 97.5$ nm (i.e., minimum channel width = 3λ, where λ is the minimum feature size) and $W_p = 243.7$ nm (which is 2.5 times W_n), power dissipation in subthreshold is 1.16 nW, whereas it is 0.30 μW in super-threshold. For the NMOS transistor width of 4 μm, st leads to 15.36 nW and ST leads to 13.05 μW power dissipation, respectively. For 90-nm technology, in subthreshold, power dissipation is 1.24 nW and 0.57 μW in super-threshold. In the two cases considered here, the NMOS transistor width is 1.38 times higher in case of 90-nm technology than 65-nm.

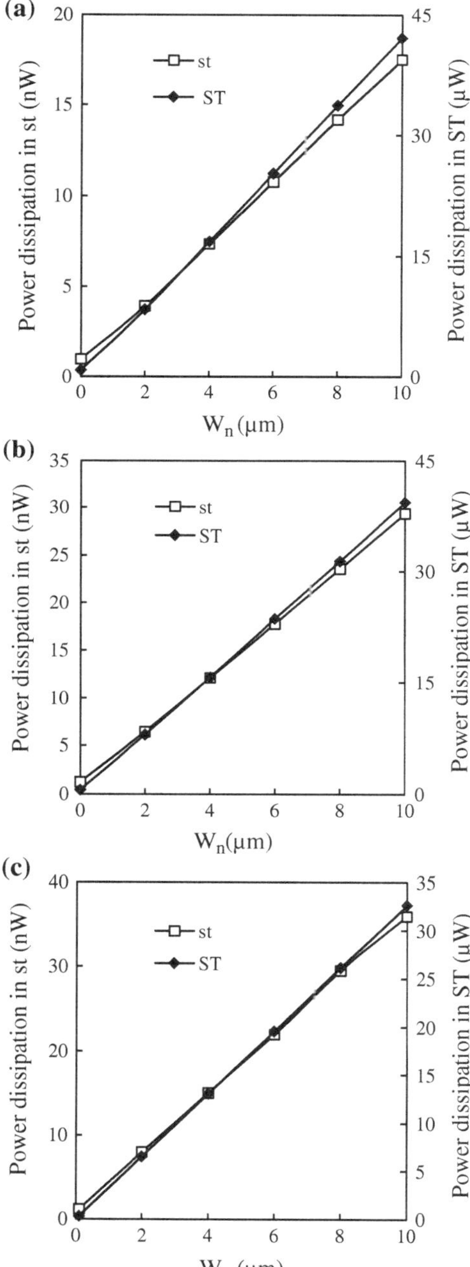

Fig. 3.6 Power dissipation variation of buffer driver interconnect load with NMOS width for **a** 130-nm, **b** 90-nm, and **c** 65-nm technology nodes

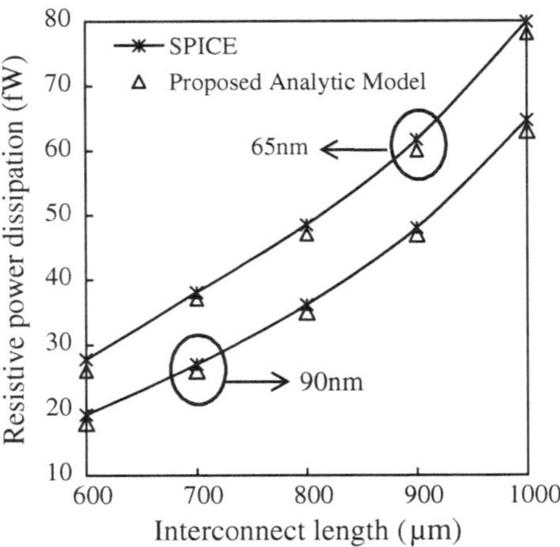

Fig. 3.7 Resistive power dissipation using SPICE and the proposed model for 90- and 65-nm technology nodes

Furthermore, power increases with buffer size. For instance, power dissipation in st is 17.75 and 22.61 nW for 90- and 65-nm technologies, respectively, for $W_n = 6$ μm. This limits the scope to use larger buffers. Hence, smaller-size buffer is preferred to contain power dissipation.

Interconnect resistance also dissipates power. However, the fraction of P_R to the total power dissipation is quite small. The variation of resistive power dissipation with L_i in subthreshold regime is shown in Fig. 3.7.

It is observed that P_R increases with L_i. The maximum percentage error between SPICE and the proposed model is 7.98 and 11.62 % for 90- and 65-nm technologies, respectively, while the average errors in the estimation of same are 6.19 and 7.32 %, respectively. Thus, the results obtained with the proposed model and SPICE are in good agreement.

The variation of NMOS resistive current with W_n is shown in Fig. 3.8. The proposed resistive current analytical model matches closely with the SPICE simulations. The maximum percentage error between SPICE and the proposed model is 7.35 % for 65-nm technology, while it is 4.60 % for 90-nm technology. The average percentage error is 3.63 % for 65-nm technology, while it is 3.51 % for 90-nm technology node.

3.4.1 Performance Metrics

Delay, power dissipation, and power-delay-product (PDP) are important performance metrics for VLSI circuits. The variation of propagation delay, power dissipation, and PDP with NMOS width for 130-nm technology node is shown in Fig. 3.9. It is observed that the proposed models replicate SPICE simulations very closely. Propagation delay decreases with NMOS width, while power dissipation increases. This is because of the fact that higher driver widths increase its current

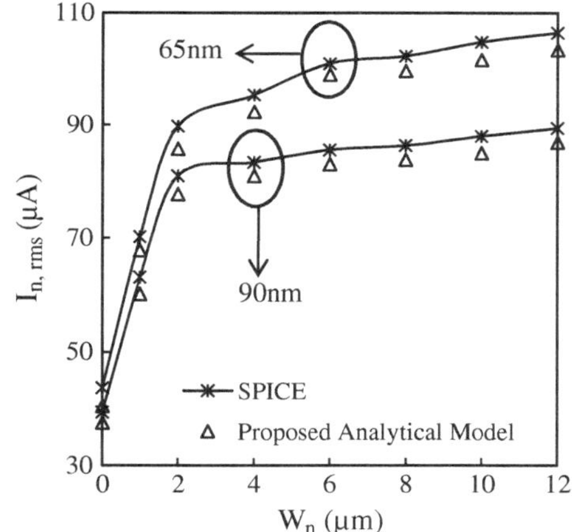

Fig. 3.8 NMOS resistive current variation using SPICE and the proposed model for 90- and 65-nm technologies with NMOS width

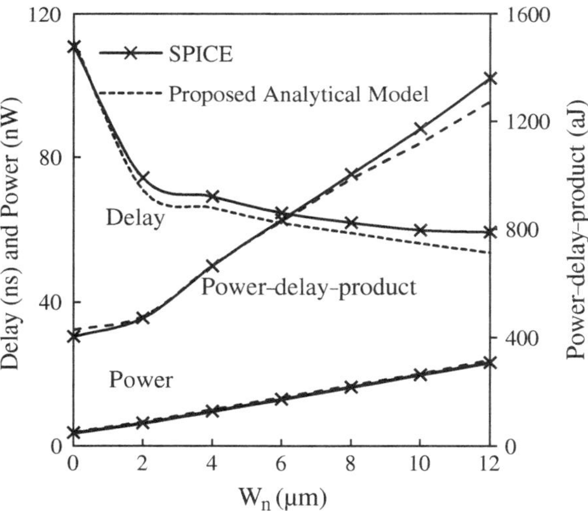

Fig. 3.9 SPICE and analytic waveforms for propagation delay, power dissipation, and power-delay-product for 130-nm technology

driving capability and hence lead to higher device performance. In this case, the average error in propagation delay does not exceed 5 %, while estimation of power dissipation leads to 4.02 % error.

PDP is a figure of merit which quantifies energy per operation in VLSI circuits. It can be considered as a quality measure for a switching device. The variation of PDP is also shown. PDP remains nearly constant up to 2 μm NMOS channel width and then increases.

The dependence of SPICE-extracted results for delay, power dissipation, and PDP on W_n for 90- and 65-nm CMOS technology nodes have been shown in Fig. 3.10a–b.

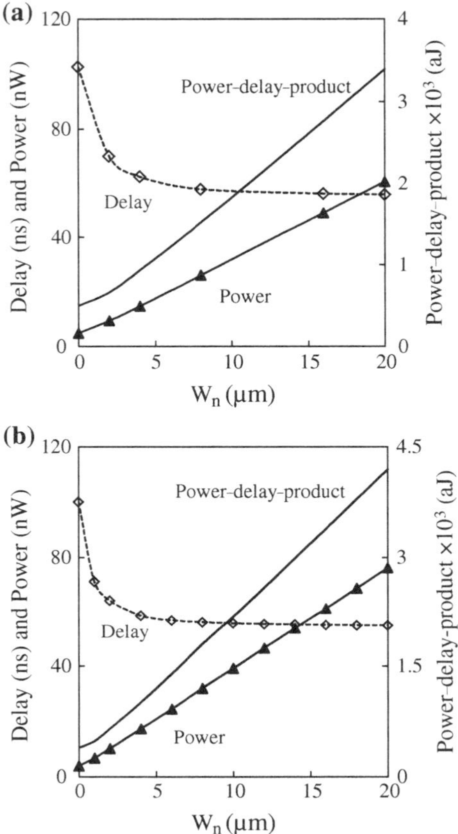

Fig. 3.10 SPICE waveforms of propagation delay, power dissipation, and power-delay-product for **a** 90-nm and **b** 65-nm technology nodes

The minimum possible transistor channel lengths are used in the respective technologies. It is found that PDP curves are characterized by a very low value of W_n which is 2 μm for all cases. Lowest PDP means lesser energy dissipation, and thereby, the pressure on the battery in the low-power devices can be reduced. Beyond this value of width, the PDP increases with increase in W_n. NMOS width of 2 μm is obtained as the optimum width for minimum PDP. At the optimal dimensions, the design shall be both power and energy efficient.

3.5 Concluding Remarks

Two sub-regions in the subthreshold regime have been characterized and their governing current–voltage equations presented. Expressions characterizing the output voltage of a CMOS buffer-driving resistive–capacitive interconnect load for rising and falling ramp inputs have been proposed. For each case of stimulation, the model gives an insight into three regions of operation of the CMOS gate. The assumption of a fast ramp signal which is widely used in the transient analysis of CMOS logic gates has been quantified in this chapter. The analytical models have also been provided for the slow ramp signal. Method has been given for determining the time at which MOSFET transits from sub-saturation to sub-linear region. The propagation delay models for both fast and slow ramp signals have been presented. The average percentage error involved in the estimation of timing with respect to SPICE is, respectively, 2.80 and 7.05 % for high-to-low and low-to-high transitions. The average error in the estimation of propagation delays is 3.89 and 8.57 % for rising and falling input ramps, respectively. The dependence of propagation delay on interconnect length has also been studied. For the three technology nodes considered, the average percentage error is within 6 %.

A new parameter, viz., subthreshold drain conductance has been introduced which is the drain or the output conductance of MOS when it operates in subthreshold. The effect of operating MOSFETs in subthreshold regime is evaluated. It is observed that subthreshold operation provides significant savings in power dissipation. This saving is over an order of magnitude. The resistive power dissipation is modeled and tracks SPICE results quite closely. The average percentage error between SPICE and the proposed model is 6.19 and 7.32 %, respectively, for 90- and 65-nm technologies, respectively. The resistive current is also modeled and is obtained within 4 % error for different NMOS widths. PDP has been characterized as a quality metric to minimize energy and optimize performance simultaneously. The PDP curves are characterized by a very low value of NMOS width which is 2 μm for all cases. Low PDP means less energy dissipation, and thereby, pressure on the battery in the low-power devices is released. The proposed model has a fairly good accuracy for deep submicron technology regime, which is presently of commercial interest in VLSI design.

To summarize, subthreshold analysis of buffer-driven interconnect has been quantified in this chapter. In practical CMOS VLSI circuits, however, interconnects are never single and exhibit coupling. The coupling between the adjacent interconnects in today's integrated circuits is large and plays a significant role in determining the signal integrity and other design issues. The following chapter addresses this issue in detail.

Chapter 4
Characterization of Dynamic Crosstalk Effect in Subthreshold Interconnects

Keywords Coupling capacitance · Dynamic crosstalk · Power-delay-product · SPICE · Timing analysis

The semiconductor electronics technology, as predicted by Moore, is toward dense, complex, and faster systems. With decreasing feature size and increasing average length of on-chip interconnections, the interconnect ground capacitance has become comparable to or larger than the input gate capacitance of the driven gate. The interconnect capacitance is therefore crucial in satisfying timing requirements. In deep submicron design, the spacing between interconnects is reduced and the thickness of the conductor is increased in order to reduce the parasitic resistance of the conductors. The coupling capacitance has therefore increased significantly and has become comparable to the interconnect capacitance.

Figure 4.1 depicts this scenario by comparing the feature sizes for two different technology regimes. The tightly coupled interconnects result in a higher probability of interaction between the electric fields of interconnects resulting in unwanted interference which causes crosstalk [224]. The crosstalk due to coupling capacitance has become extremely important in technologies below 0.18 μm. The coupling capacitance increases the propagation delay and power dissipation and alters the waveform shape of the output voltage signal [225]. The delay impact due to crosstalk is extremely important, since regular static timing analysis considers all coupled interconnect lines to be quiet, which is seldom the case. The coupling capacitance is therefore an important design parameter in evaluating the signal integrity of interconnects in a CMOS VLSI chip. Consequently, the effects of the global interconnect impedance parameters particularly on delay is of great concern for the VLSI circuit designers.

Crosstalk in coupled lines can be broadly divided into two categories, viz., (i) functional crosstalk and (ii) dynamic crosstalk. Under functional crosstalk, overshoots and undershoots are experienced on the victim (quiet) line because of switching activity on the aggressor (active) lines. Overshoots and undershoots may cause current to flow through the substrate, possibly corrupting data in dynamic logic circuits [226]. Under dynamic crosstalk, noise is experienced when aggressor and victim lines switch simultaneously. Since it is common to encounter dynamic crosstalk in practice, its analysis is as important as that of functional crosstalk noise.

R. Dhiman and R. Chandel, *Compact Models and Performance Investigations for Subthreshold Interconnects*, Energy Systems in Electrical Engineering,
DOI 10.1007/978-81-322-2132-6_4

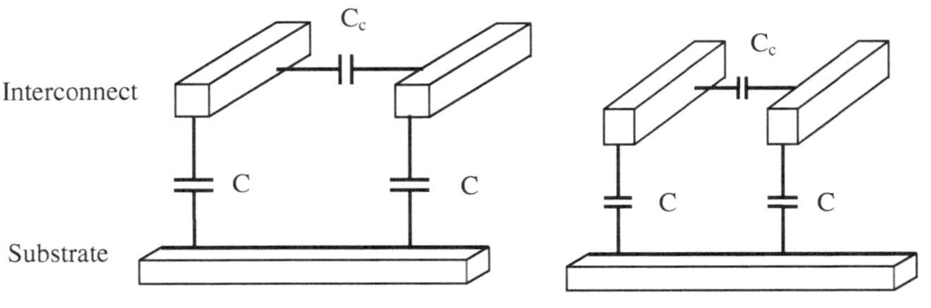

Fig. 4.1 Comparison of coupling and substrate capacitances

This dynamic form of coupling causes a change in the signal propagation delay and thus impacts the critical issue of timing. An accurate model for transient analysis of dynamic crosstalk is therefore important. The present chapter is an attempt in this direction and mainly focuses on the crosstalk analysis and delay estimation in simultaneously switching coupled scenario for subthreshold interconnect circuits. To obtain the solution analytically, subthreshold current model of CMOS buffer has been taken. In-phase and out-of-phase input switching conditions have been considered. An analytical framework to characterize output voltages and propagation delays in subthreshold regime for CMOS buffer-driven coupled VLSI interconnects has been developed. These delay relationships have been developed for both fast and slow ramp input signals.

4.1 The Output Voltage of Each CMOS Inverter

The proposed analytical approach considers CMOS gates driving two capacitively coupled lines. Subthreshold model of a MOS transistor is used to analyze a CMOS driver. This is combined with coupled resistive–capacitive model of interconnect to derive analytical closed-form expressions. Interconnect is modeled as lumped resistive–capacitive in order to emphasize the nonlinear behavior of the MOS devices. Such a representation of the composite model where two capacitively coupled lines each driven by CMOS inverter (*Inv*) has been shown in Fig. 4.2a. The equivalent circuit for the same is shown in Fig. 4.2b. R_1 (R_2) is the parasitic interconnect resistance, C_1 (C_2) is the intrinsic capacitance and includes the interconnect ground capacitance and the input gate capacitance of *Inv*3 (*Inv*4), and C_c is the coupling capacitance between the wires. Inverters 1 and 2, viz., *Inv*1 and *Inv*2, are the aggressor and victim drivers, respectively.

The equivalent circuit model also shows the related current directions as shown in Fig. 4.2b. The effects of the coupling capacitance on the transient response of coupled interconnects depend upon the switching activity. Two switching conditions are considered as follows:

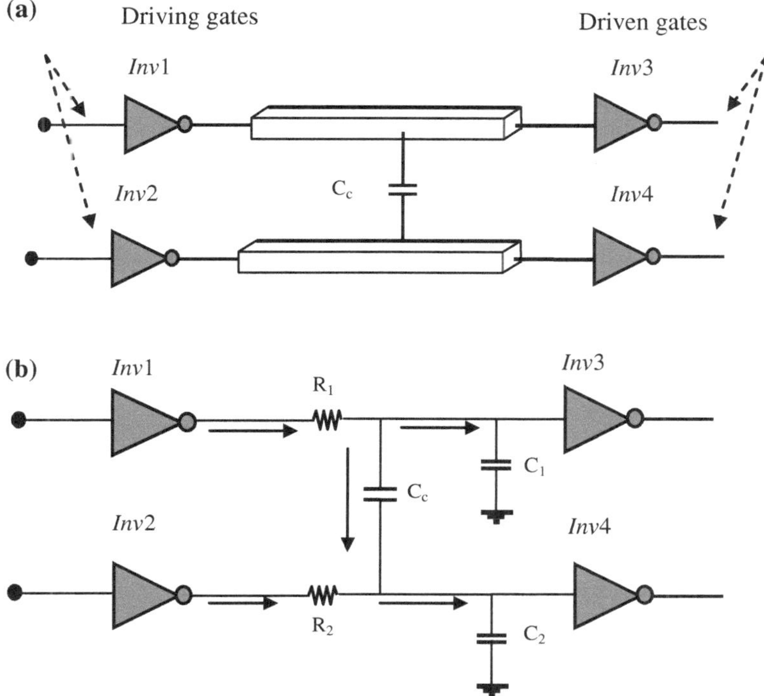

Fig. 4.2 **a** Circuit model of buffer (inverter)-driven capacitively coupled interconnect lines. **b** Equivalent circuit of two capacitively coupled resistive–capacitive interconnections driven by CMOS buffers

I V_{in1} and V_{in2} switching from low-to-high. Thus, V_{in1} (input to *Inv*1) and V_{in2} (input to *Inv*2) are switching in the same direction or in-phase.

II V_{in1} is switching from low-to-high, and V_{in2} switches from high-to-low. Thus, V_{in1} and V_{in2} are switching out-of-phase.

On this basis, expressions for the dynamic crosstalk have been developed and analyses of in-phase and out-of-phase switching presented. In the foregoing analysis, it is assumed that both inverters are triggered at the same time and have equal rise (fall) times.

4.2 In-Phase Switching

The in-phase switching is an optimistic condition in terms of the effect of the coupling capacitance on the propagation delay of each CMOS inverter. For a two-line coupled system, it is assumed that inputs of both inverters transition from low-to-high.

MN1 and MN2 are the active transistors in each inverter. The PMOS transistors are neglected under the assumption of fast ramp input signal. An assumption of fast ramp input signal permits the condition that both transistors, i.e., MN1 and MN2, operate in sub-saturation even after the completion of input transition.

In this section, analytical expressions governing the output voltage and propagation delay of each CMOS inverter are presented.

The input signal driving both CMOS buffers is characterized by

$$\begin{array}{ll} V_{\text{in1}} = V_{\text{in2}} = V_{\text{DD}}\frac{t}{\tau_r} & 0 \le t \le \tau_r \\ V_{\text{in1}} = V_{\text{in2}} = V_{\text{DD}} & t > \tau_r \end{array} \tag{4.1}$$

The differential equations governing the output voltage of each MOS transistor shown in Fig. 4.3 are given by

$$-\frac{dV_1}{dt} = \frac{(C_2 + C_c)}{C_1C_2 + C_c(C_1 + C_2)} I_{n1} + \frac{C_c}{C_1C_2 + C_c(C_1 + C_2)} I_{n2} + R_1\frac{dI_{n1}}{dt} \tag{4.2}$$

$$-\frac{dV_2}{dt} = \frac{(C_1 + C_c)}{C_1C_2 + C_c(C_1 + C_2)} I_{n2} + \frac{C_c}{C_1C_2 + C_c(C_1 + C_2)} I_{n1} + R_2\frac{dI_{n2}}{dt} \tag{4.3}$$

I_{n1} and I_{n2} are the currents flowing across MN1 and MN2, respectively. For the rising ramp, MOS transistors operate in two sub-regions, viz., sub-saturation and sub-linear. In order to obtain the output voltages V_1 and V_2 of *Inv*1 and *Inv*2, respectively, four operating regions have been identified and discussed below.

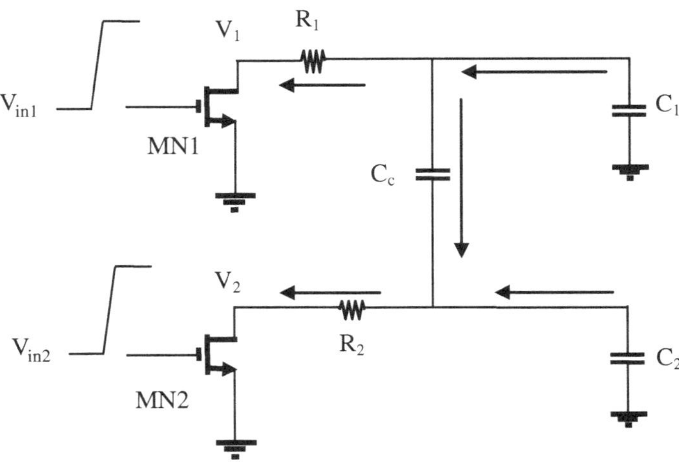

Fig. 4.3 Equivalent circuit for buffer-driven coupled interconnects simultaneously switching in-phase for low-to-high input transitions

Region-1 ($0 \leq t \leq \tau_r$): The initial values of V_1 and V_2 both are equal to V_{DD}. The currents across MN1 and MN2 in the sub-saturation region are given by

$$I_{n1} = B_{n1}\exp\left[\frac{V_{in1} - V_{DD}}{\eta_n U_{th}}\right] = B_{n1}\exp\left[\frac{V_{DD}\frac{t}{\tau_r} - V_{DD}}{\eta_n U_{th}}\right] \tag{4.4}$$

$$I_{n2} = B_{n2}\exp\left[\frac{V_{in2} - V_{DD}}{\eta_n U_{th}}\right] = B_{n2}\exp\left[\frac{V_{DD}\frac{t}{\tau_r} - V_{DD}}{\eta_n U_{th}}\right] \tag{4.5}$$

The output voltages obtained are

$$V_1 = V_{DD} - \gamma_{21}\frac{\tau_r\eta_n U_{th}}{V_{DD}}\left[e^{\frac{V_{DD}\frac{t}{\tau_r} - V_{DD}}{\eta_n U_{th}}} - e^{-\frac{V_{DD}}{\eta_n U_{th}}}\right] - R_1 B_{n1}\left[e^{\frac{V_{DD}\frac{t}{\tau_r} - V_{DD}}{\eta_n U_{th}}} - e^{-\frac{V_{DD}}{\eta_n U_{th}}}\right] \tag{4.6}$$

$$V_2 = V_{DD} - \gamma_{22}\frac{\tau_r\eta_n U_{th}}{V_{DD}}\left[e^{\frac{V_{DD}\frac{t}{\tau_r} - V_{DD}}{\eta_n U_{th}}} - e^{-\frac{V_{DD}}{\eta_n U_{th}}}\right] - R_2 B_{n2}\left[e^{\frac{V_{DD}\frac{t}{\tau_r} - V_{DD}}{\eta_n U_{th}}} - e^{-\frac{V_{DD}}{\eta_n U_{th}}}\right] \tag{4.7}$$

The various constants in (4.6) and (4.7) are defined as

$$\gamma_{21} = \frac{(C_2 + C_c)B_{n1} + C_c B_{n2}}{C_1 C_2 + C_c(C_1 + C_2)} \tag{4.8}$$

$$\gamma_{22} = \frac{(C_1 + C_c)B_{n2} + C_c B_{n1}}{C_1 C_2 + C_c(C_1 + C_2)} \tag{4.9}$$

The effect of coupling can be seen in Eqs. (4.6) and (4.7). Both γ_{21} and γ_{22} include the effect of coupling capacitance. The coupling capacitance thus affects the propagation delay. This delay uncertainty in the propagation delay can be eliminated if MN1 and MN2 both have the same ratio of output current drive (B_{n1}/B_{n2}) to the corresponding intrinsic load capacitances (C_1/C_2). Under such conditions, the coupling capacitance has no effect on the output voltage waveforms V_1 and V_2. However, this condition is difficult to be realized in practical CMOS VLSI circuits. This is owing to the different geometric sizes of MOS transistors ($B_{n1} \neq B_{n2}$), different interconnect geometric parameters and different gate-to-source capacitances ($C_1 \neq C_2$) of the fan-out logic gates. Thus, coupling capacitance always affects the output voltage, and hence, timing analysis under such switching environment becomes necessary. At time t equal to τ_r, the output voltages obtained are

$$V_1(\tau_r) = V_{DD} - \gamma_{21}\frac{\tau_r\eta_n U_{th}}{V_{DD}}\left[1 - e^{-\frac{V_{DD}}{\eta_n U_{th}}}\right] - R_1 B_{n1}\left[1 - e^{-\frac{V_{DD}}{\eta_n U_{th}}}\right] \tag{4.10}$$

$$V_2(\tau_r) = V_{DD} - \gamma_{22}\frac{\tau_r\eta_n U_{th}}{V_{DD}}\left[1 - e^{-\frac{V_{DD}}{\eta_n U_{th}}}\right] - R_2 B_{n2}\left[1 - e^{-\frac{V_{DD}}{\eta_n U_{th}}}\right] \tag{4.11}$$

Region-2 ($\tau_r \leq t \leq \tau_{\text{nsat}}^1$): The drain-to-source currents of MN1 and MN2 are constants and given by $I_1 = B_{\text{n1}}$ and $I_2 = B_{\text{n2}}$. The output voltages in this region are based on the condition at $t = \tau_{\text{r}}$ and are

$$V_1 = V_1(\tau_{\text{r}}) - \gamma_{21}(t - \tau_{\text{r}}) \tag{4.12}$$

$$V_2 = V_2(\tau_{\text{r}}) - \gamma_{22}(t - \tau_{\text{r}}) \tag{4.13}$$

Region-3 ($\tau_{\text{nsat}}^1 \leq t \leq \tau_{\text{nsat}}^2$): Depending on the geometric size of MOS transistors, it is possible that MN1 leaves the sub-saturation region and enters into sub-linear region, while MN2 continues to operate in the sub-saturation. MN1 and MN2 make transition into the sub-linear region of their characteristics at duration times τ_{nsat}^1 and τ_{nsat}^2, respectively. Further, it is assumed that MN1 leaves the sub-saturation region first, i.e., $\tau_{\text{nsat}}^1 < \tau_{\text{nsat}}^2$. In this case, the drain-to-source current of MN1 is characterized by

$$I_1 = \gamma_{\text{n1}} V_1 \tag{4.14}$$

γ_{n1} is the output conductance of MN1 in the sub-linear region. The output voltages obtained in this region are

$$V_1 = -V_{\text{a}} - \left[V_1\left(\tau_{\text{nsat}}^1\right) + V_{\text{a}}\right] \mathrm{e}^{-\alpha_{\text{n1}}\left(t - \tau_{\text{nsat}}^1\right)} \tag{4.15}$$

$$V_2 = V_2\left(\tau_{\text{nsat}}^1\right) - V_{\text{b}} - \frac{B_{\text{n2}}}{C_2 + C_{\text{c}}}\left(t - \tau_{\text{nsat}}^1\right) \tag{4.16}$$

where

$$V_{\text{a}} = \frac{C_{\text{c}}}{(C_2 + C_{\text{c}})\gamma_{\text{n1}}} B_{\text{n2}} \tag{4.17}$$

$$\alpha_{\text{n1}} = \frac{(C_2 + C_{\text{c}})}{[C_1 C_2 + C_{\text{c}}(C_1 + C_2)](1 + \gamma_{\text{n1}} R_1)} \gamma_{\text{n1}} \tag{4.18}$$

$$V_{\text{b}} = \frac{C_{\text{c}}}{(C_2 + C_{\text{c}})}(1 + \gamma_{\text{n1}} R_1)\left[V_1\left(\tau_{\text{nsat}}^1\right) + V_{\text{a}}\right]\left(1 - \mathrm{e}^{-\alpha_{\text{n1}}\left(t - \tau_{\text{nsat}}^1\right)}\right) \tag{4.19}$$

Region-4 ($t \geq \tau_{\text{nsat}}^2$): After τ_{nsat}^2, both of the NMOS transistors operate in the sub-linear region. The differential equations governing the output voltage of each MOS transistor are given by

$$-(C_1 + C_{\text{c}})(1 + \gamma_{\text{n1}} R_1)\frac{\mathrm{d}V_1}{\mathrm{d}t} + C_{\text{c}}(1 + \gamma_{\text{n2}} R_2)\frac{\mathrm{d}V_2}{\mathrm{d}t} = \gamma_{\text{n1}} V_1 \tag{4.20}$$

$$-(C_2 + C_{\text{c}})(1 + \gamma_{\text{n2}} R_2)\frac{\mathrm{d}V_2}{\mathrm{d}t} + C_{\text{c}}(1 + \gamma_{\text{n1}} R_1)\frac{\mathrm{d}V_1}{\mathrm{d}t} = \gamma_{\text{n2}} V_2 \tag{4.21}$$

Here, γ_{n2} is the output conductance of MN2 in the sub-linear region. Analytical expressions characterizing the output voltage of each CMOS inverter obtained are

$$\begin{aligned} V_1 &= \frac{1}{2} V_1\left(\tau_{\text{nsat}}^2\right)\left[\mathrm{e}^{\frac{\chi-(a_1+b_1)}{2}\left(t-\tau_{\text{nsat}}^2\right)}\left(1+\frac{b_1-a_1}{\chi}\right)+\mathrm{e}^{-\frac{\chi+(a_1+b_1)}{2}\left(t-\tau_{\text{nsat}}^2\right)}\left(1-\frac{b_1-a_1}{\chi}\right)\right] \\ &\quad -\frac{a_2}{\chi} V_2\left(\tau_{\text{nsat}}^2\right)\left[\mathrm{e}^{\frac{\chi-(a_1+b_1)}{2}\left(t-\tau_{\text{nsat}}^2\right)}-\mathrm{e}^{-\frac{\chi+(a_1+b_1)}{2}\left(t-\tau_{\text{nsat}}^2\right)}\right] \end{aligned} \tag{4.22}$$

$$\begin{aligned} V_2 &= \frac{1}{2} V_2\left(\tau_{\text{nsat}}^2\right)\left[\mathrm{e}^{\frac{\chi-(a_1+b_1)}{2}\left(t-\tau_{\text{nsat}}^2\right)}\left(1+\frac{a_1-b_1}{\chi}\right)+\mathrm{e}^{-\frac{\chi+(a_1+b_1)}{2}\left(t-\tau_{\text{nsat}}^2\right)}\left(1-\frac{a_1-b_1}{\chi}\right)\right] \\ &\quad -\frac{b_2}{\chi} V_1\left(\tau_{\text{nsat}}^2\right)\left[\mathrm{e}^{\frac{\chi-(a_1+b_1)}{2}\left(t-\tau_{\text{nsat}}^2\right)}-\mathrm{e}^{-\frac{\chi+(a_1+b_1)}{2}\left(t-\tau_{\text{nsat}}^2\right)}\right] \end{aligned} \tag{4.23}$$

$V_1\left(\tau_{\text{nsat}}^2\right)$ and $V_2\left(\tau_{\text{nsat}}^2\right)$ are initial values of V_1 and V_2 at $t = \tau_{\text{nsat}}^2$. The various constants a_1, a_2, b_1, b_2, and χ are defined as

$$a_1 = \frac{(C_2 + C_c)}{C_1C_2 + C_c(C_1 + C_2)}\gamma_{n1} \tag{4.24}$$

$$a_2 = \frac{C_c}{C_1C_2 + C_c(C_1 + C_2)}\gamma_{n2} \tag{4.25}$$

$$b_1 = \frac{(C_1 + C_c)}{C_1C_2 + C_c(C_1 + C_2)}\gamma_{n2} \tag{4.26}$$

$$b_2 = \frac{C_c}{C_1C_2 + C_c(C_1 + C_2)}\gamma_{n1} \tag{4.27}$$

$$\chi = \sqrt{(\alpha_1 - \beta_1)^2 + 4\alpha_2\beta_2} \tag{4.28}$$

4.2.1 Propagation Delay for Fast Ramp

The high-to-low propagation delays $t_{P_{HL1}}$ and $t_{P_{HL2}}$ of MN1 and MN2, respectively, are computed based on (4.12) and (4.13). At $t = t_{P_{HL1}}$ and $t = t_{P_{HL2}}$, the output voltages V_1 and V_2 both are equal to $0.5V_{DD}$, i.e.,

$$V_1(\tau_r) - \gamma_{21}(t_{P_{HL1}} - \tau_r) = 0.5V_{DD} \tag{4.29}$$

$$V_2(\tau_r) - \gamma_{22}(t_{P_{HL2}} - \tau_r) = 0.5V_{DD} \tag{4.30}$$

Simplification of (4.29) and (4.30) gives

$$t_{P_{HL1}} = \tau_r + \frac{0.5V_{DD} - R_1 B_{n1}\left(1 - e^{-\frac{V_{DD}}{\eta_n U_{th}}}\right)}{\gamma_{21}} - \frac{\tau_r \eta_n U_{th}\left(1 - e^{-\frac{V_{DD}}{\eta_n U_{th}}}\right)}{V_{DD}} \tag{4.31}$$

$$t_{P_{HL2}} = \tau_r + \frac{0.5V_{DD} - R_2 B_{n2}\left(1 - e^{-\frac{V_{DD}}{\eta_n U_{th}}}\right)}{\gamma_{22}} - \frac{\tau_r \eta_n U_{th}\left(1 - e^{-\frac{V_{DD}}{\eta_n U_{th}}}\right)}{V_{DD}} \tag{4.32}$$

It is seen from (4.31) and (4.32) that the propagation delay time depends upon the intrinsic load capacitances, the coupling capacitance, and the size of the active transistor in each CMOS inverter. The propagation delay is the average of the high-to-low and low-to-high propagation delays. These are denoted by τ_{p1} and τ_{p2} for the aggressor and victim buffers, respectively.

τ^1_{nsat} and τ^2_{nsat} are calculated based on the boundary condition defined as

$$V_1(\tau_r) - \gamma_{21}\left(\tau^1_{nsat} - \tau_r\right) = 4U_{th} \tag{4.33}$$

$$V_2\left(\tau^2_{nsat}\right) - V_b - \frac{B_{n2}}{C_2 + C_c}\left(\tau^2_{nsat} - \tau^1_{nsat}\right) = 4U_{th} \tag{4.34}$$

Equation (4.33) is solved to yield

$$\tau^1_{nsat} = \tau_r + \frac{V_1(\tau_r) - 4U_{th}}{\gamma_{21}} \tag{4.35}$$

τ^2_{nsat} can be computed using Eq. (4.34) and Newton–Raphson numeric solver.

4.2.2 Propagation Delay for Slow Ramp

The model proposed in this section is based on the assumption of a slow ramp input. The output voltages of coupled buffers in the time interval $0 \le t \le \tau^1_{nsat}$ are essentially similar to (4.6) and (4.7). In the time interval $\tau^1_{nsat} \le t \le \tau^2_{nsat}$, MN2 continues to operate in the sub-saturation region. However, MN2 makes transition to sub-linear region of its characteristics at $t = \tau^2_{nsat}$. The output voltages of each CMOS inverter obtained are given by

$$V_1 = V_1\left(\tau^1_{nsat}\right)e^{-\alpha_{n1}\left(t-\tau^1_{nsat}\right)} - \frac{C_c B_{n2}\tau_r \eta_n U_{th} e^{-\frac{V_{DD}}{\eta_n U_{th}}}}{V_{DD}[C_1 C_2 + C_c(C_1 + C_2)] + \tau_r \eta_n U_{th}(C_2 + C_c)\gamma_{n1}}\left[e^{\frac{V_{DD}\left(t-\tau^1_{nsat}\right)}{\tau_r \eta_n U_{th}}} - e^{-\alpha_{n1}\left(t-\tau^1_{nsat}\right)}\right] \tag{4.36}$$

$$V_2 = V_2(\tau_{\text{nsat}}^1) - R_2 B_{\text{n2}} - \frac{B_{\text{n2}}\tau_{\text{r}}\eta_{\text{n}}U_{\text{th}}(C_1 + C_{\text{c}})}{V_{\text{DD}}[C_1C_2 + C_{\text{c}}(C_1 + C_2)]} e^{-\frac{V_{\text{DD}}}{\eta_{\text{n}}U_{\text{th}}}} \left[e^{V_{\text{DD}}\frac{t-\tau_{\text{nsat}}^1}{\tau_{\text{r}}}} - 1 \right]$$
$$- \frac{C_{\text{c}}}{C_1C_2 + C_{\text{c}}(C_1 + C_2)} \gamma_{\text{n1}} \int_{t-\tau_{\text{nsat}}^1}^{t} V_1 \text{d}t \quad (4.37)$$

The high-to-low propagation delays $t_{0.5}^1$ and $t_{0.5}^2$ for slow ramp signal can be obtained by applying Newton–Raphson numeric solver for (4.36) and (4.37) relations.

4.3 Out-of-Phase Switching

The out-of-phase transition is a pessimistic condition in terms of the effect of the coupling capacitance on the propagation delays of CMOS inverters [221]. In this case, it is assumed that input to *Inv*1 is switching from low-to-high and input to *Inv*1 is switching from high-to-low as shown in Fig. 4.4. MN1 and MP2 are the active transistors in each inverter for the considered input conditions. The initial values of V_1 and V_2 are V_{DD} and ground, respectively. The directions of the currents are also shown. An assumption of fast ramp input signal is considered.

The shape of input signals driving *Inv*1 and *Inv*2 is

$$V_{\text{in1}} = V_{\text{in2}} = V_{\text{DD}} \frac{t}{\tau_{\text{r}}} \quad \text{for} \ \ 0 \le t \le \tau_{\text{r}} \quad (4.38)$$

$$V_{\text{in2}} = V_{\text{DD}} \left(1 - \frac{t}{\tau_{\text{r}}}\right) \quad \text{for} \ \ 0 \le t \le \tau_{\text{r}} \quad (4.39)$$

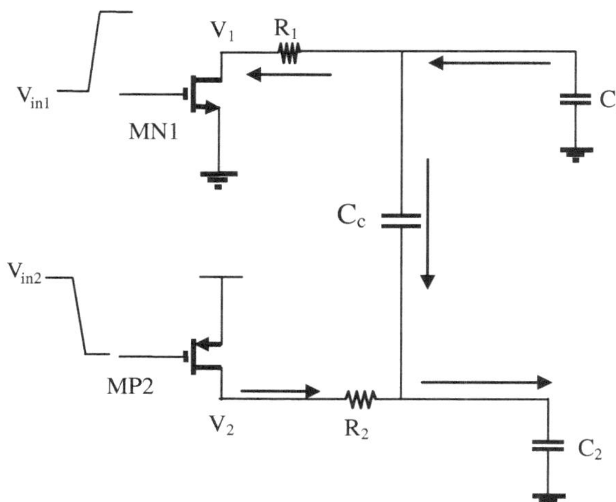

Fig. 4.4 Equivalent circuit for aggressor input switching from low-to-high and victim input switching from high-to-low

The differential equations governing the output voltage of each MOS transistor shown in Fig. 4.4 are given by

$$-\frac{dV_1}{dt}=\frac{(C_2+C_c)}{C_1C_2+C_c(C_1+C_2)}I_{n1}-\frac{C_c}{C_1C_2+C_c(C_1+C_2)}I_{p2}+R_1\frac{dI_{n1}}{dt} \quad (4.40)$$

$$-\frac{dV_2}{dt}=\frac{C_c}{C_1C_2+C_c(C_1+C_2)}I_{n1}-\frac{(C_1+C_c)}{C_1C_2+C_c(C_1+C_2)}I_{p2}-R_2\frac{dI_{p2}}{dt} \quad (4.41)$$

Here, I_{p2} is the current that flows across MP2. For the opposite switching condition with fast ramp input, MOS transistors operate in different regions over different intervals of time. Following the similar procedure as for in-phase switching, analytical expressions characterizing the output voltage of each CMOS inverter for out-of-phase switching condition are derived.

Region-1 $(0\le t\le \tau_r)$: In region-1, MN1 of the aggressor driver and MP2 of victim driver operate in the sub-saturation regions. As MN1 is in sub-saturation, the current across MN1 is given by (4.4). The current across MP2 is given by

$$I_{p2}=B_{p2}e^{\frac{V_{DD}\frac{t}{\tau_r}-V_{DD}}{\eta_p U_{th}}} \quad (4.42)$$

B_{p2} is the source-to-drain current of MP2 when $V_{in2}=V_{DD}$ and η_p is its sub-threshold slope factor. The output voltages V_1 and V_2 of *Inv*1 and *Inv*2, respectively, obtained are given by

$$V_1=V_{DD}-\frac{(C_2+C_c)V_{n,1}-C_cV_{p,2}}{C_1C_2+C_c(C_1+C_2)}-R_1B_{n1}\left[e^{\frac{V_{DD}\frac{t}{\tau_r}-V_{DD}}{\eta_n U_{th}}}-e^{-\frac{V_{DD}}{\eta_n U_{th}}}\right] \quad (4.43)$$

$$V_2=\frac{(C_1+C_c)V_{p,2}-C_cV_{n,1}}{C_1C_2+C_c(C_1+C_2)}+R_2B_{p2}\left[e^{\frac{V_{DD}\frac{t}{\tau_r}-V_{DD}}{\eta_p U_{th}}}-e^{-\frac{V_{DD}}{\eta_p U_{th}}}\right] \quad (4.44)$$

where

$$V_{n,1}=\frac{B_{n1}\tau_r\eta_n U_{th}}{V_{DD}}\left[e^{\frac{V_{DD}\frac{t}{\tau_r}-V_{DD}}{\eta_n U_{th}}}-e^{-\frac{V_{DD}}{\eta_n U_{th}}}\right] \quad (4.45)$$

$$V_{p,2}=\frac{B_{p2}\tau_r\eta_p U_{th}}{V_{DD}}\left[e^{\frac{V_{DD}\frac{t}{\tau_r}-V_{DD}}{\eta_p U_{th}}}-e^{-\frac{V_{DD}}{\eta_p U_{th}}}\right] \quad (4.46)$$

The effect of coupling can be observed in (4.43) and (4.44). Coupling affects V_1 and V_2 through $V_{n,1}$ and $V_{p,2}$, respectively. It can be observed that the presence of the coupling term $V_{p,2}$ in (4.43) tends to decrease V_1 slowly, while the coupling component $V_{n,1}$ causes V_2 to increase slowly in (4.44).

Region-2 ($\tau_r \leq t \leq \tau_{nsat}^1$): After τ_r, both inputs attain fixed value equal to V_{DD} and ground, respectively. MN1 and MP2 continue to operate in the sub-saturation. For this duration, the voltages at the output of both buffers are given by

$$V_1 = V_1(\tau_r) - \frac{(C_2 + C_c)B_{n1} - C_c B_{p2}}{C_1 C_2 + C_c(C_1 + C_2)}(t - \tau_r) \tag{4.47}$$

$$V_2 = V_2(\tau_r) + \frac{(C_1 + C_c)B_{p2} - C_c B_{n1}}{C_1 C_2 + C_c(C_1 + C_2)}(t - \tau_r) \tag{4.48}$$

Region-3 ($\tau_{nsat}^1 \leq t \leq \tau_{psat}^2$): MN1 and MP2 may leave sub-saturation region at different time durations if both transistors have unequal output conductances. MP2 makes transition in sub-linear region at $t = \tau_{psat}^2$. In region-3, therefore, MN1 operates in the sub-linear region, while MP2 continues to remain in the sub-saturation region. The relationship between V_1 and V_2 is given by

$$V_1 = V_{1,a} - \left[-V_1\left(\tau_{nsat}^1\right) + V_{1,a}\right] e^{-\alpha_{n1}\left(t - \tau_{nsat}^1\right)} \tag{4.49}$$

$$V_2 = V_2\left(\tau_{nsat}^1\right) - V_{1,b} + \frac{B_{p2}}{(C_2 + C_c)}\left(t - \tau_{nsat}^1\right) \tag{4.50}$$

where

$$V_{1,a} = \frac{C_c}{\gamma_{n1}(C_2 + C_c)} B_{p2} \tag{4.51}$$

$$V_{1,b} = \frac{C_c}{C_2 + C_c}(1 + \gamma_{n1}R_1)\left[V_{1,a} - V_1\left(\tau_{nsat}^1\right)\right]\left(1 - e^{-\alpha_{n1}\left(t - \tau_{nsat}^1\right)}\right) \tag{4.52}$$

Region-4 ($t > \tau_{psat}^2$): MN1 and MP2 operate in the sub-linear region. The differential equations governing the output voltage of each MOS transistor are given by

$$-(C_1 + C_c)(1 + \gamma_{n1}R_1)\frac{dV_1}{dt} + C_c(1 + \gamma_{p2}R_2)\frac{dV_2}{dt} = \gamma_{n1}V_1 \tag{4.53}$$

$$-C_c(1 + \gamma_{n1}R_1)\frac{dV_1}{dt} + (C_2 + C_c)(1 + \gamma_{p2}R_2)\frac{dV_1}{dt} = \gamma_{p2}V_2 \tag{4.54}$$

These coupled differential equations are solved, and the solution of these coupled differential equations is given as

$$V_1 = \frac{1}{2\chi} e^{\frac{[\chi - (a_1 - b_1)]}{2}\left(t - \tau_{psat}^2\right)} \left[\begin{array}{c} V_1\left(\tau_{psat}^2\right)(\chi - a_1 + b_1) - 2a_2 V_2\left(\tau_{psat}^2\right) - \frac{1}{(a_1 b_1 - a_2 b_2)} \\ \left\{a_3\left(b_1^2 - a_1 b_1 + b_1\chi + 2a_2 b_2\right) - (\chi + a_1 + b_1)a_2 b_3\right\} \end{array}\right] + \frac{1}{a_1 b_1 - a_2 b_2}(a_3 b_1 - a_2 b_3) \tag{4.55}$$

$$V_2 = \frac{1}{2\chi} e^{\frac{[\chi-(a_1+b_1)]}{2}\left(t-\tau_{\text{psat}}^2\right)} \begin{bmatrix} \left\{V_2\left(\tau_{\text{psat}}^2\right)(\chi+a_1-b_1)-2\beta_2 V_1\left(\tau_{\text{psat}}^2\right)\right\}+\frac{1}{(a_1b_1-a_2b_2)} \\ \left\{a_3b_2(\chi+a_1+b_1)-\left(a_1^2b_3-a_1b_1b_3+2a_2b_2b_3+a_1b_3\chi\right)\right\} \end{bmatrix} + \frac{1}{a_1b_1-a_2b_2}(a_1b_3-a_3b_2) \quad (4.55)$$

where

$$a_3 = a_2 V_{\text{DD}} \quad (4.57)$$

$$b_3 = b_2 V_{\text{DD}} \quad (4.58)$$

$V_1\left(\tau_{\text{psat}}^2\right)$, and $V_2\left(\tau_{\text{psat}}^2\right)$ are initial values of V_1 and V_2 at the time τ_{psat}^2. Here, γ_{p_2} is the output conductance of MP2 in the sub-linear region.

4.3.1 Propagation Delay for Fast Ramp

The high-to-low ($t_{\text{P}_{\text{HL1}}}$) and low-to-high ($t_{\text{P}_{\text{LH2}}}$) propagation delays, respectively, of MN1 and MP2 are computed based on Eqs. (4.47) and (4.48). At $t = t_{\text{P}_{\text{LH1}}}$ and t $=t_{\text{P}_{\text{LH2}}}$,

$$V_1(\tau_r) - \frac{(C_2+C_c)B_{\text{n1}} - C_cB_{\text{p2}}}{C_1C_2 + C_c(C_1+C_2)}\left(t_{\text{P}_{\text{HL1}}} - \tau_r\right) = 0.5V_{\text{DD}} \quad (4.59)$$

$$V_2(\tau_r) + \frac{(C_1+C_c)B_{p2} - C_cB_{\text{n1}}}{C_1C_2 + C_c(C_1+C_2)}\left(t_{\text{P}_{\text{LH2}}} - \tau_r\right) = 0.5V_{\text{DD}} \quad (4.60)$$

Simplifying (4.59) and (4.60) gives

$$t_{\text{P}_{\text{HL1}}} = \tau_r + \frac{[V_1(\tau_r) - 0.5V_{\text{DD}}][C_1C_2 + C_c(C_1+C_2)]}{B_{\text{n1}}(C_2+C_c) - C_cB_{\text{p2}}} \quad (4.61)$$

$$t_{\text{P}_{\text{LH2}}} = \tau_r + \frac{[0.5V_{\text{DD}} - V_2(\tau_r)][C_1C_2 + C_c(C_1+C_2)]}{(C_1+C_c)B_{\text{p2}} - C_cB_{\text{n1}}} \quad (4.62)$$

The average propagation delay for the aggressor and victim buffers is obtained in a similar fashion as explained in Sect. 4.2.1. The dependence of delays on the coupling capacitance and the load capacitance can again be seen in (4.61) and (4.62). τ_{nsat}^1 is calculated based on (4.47) and is given by

$$\tau_{\text{nsat}}^1 = \tau_r + \left[\{V_1(\tau_r) - 4U_{\text{th}}\} \times \frac{C_1C_2 + C_c(C_1+C_2)}{(C_2+C_c)B_{\text{n1}} - C_cB_{\text{p2}}}\right] \quad (4.63)$$

τ^2_{psat} is calculated using Newton–Raphson numerical solver depending upon the condition

$$V_2\left(\tau^1_{\text{nsat}}\right) - V_{1,\text{b}} + \frac{B_{\text{p2}}}{(C_2 + C_\text{c})}\left(\tau^2_{\text{psat}} - \tau^1_{\text{nsat}}\right) = V_{\text{DD}} - 4U_{\text{th}} \quad (4.64)$$

4.4 Comparison with Simulation Results

The proposed models are validated using SPICE simulations. The rise time of the input ramp taken is 0.1 μs. For CMOS driver, data of PTM 65-nm, 0.36 V, and Level-54 are used throughout this chapter. The coupled interconnects have coupling length equal to 5 mm, while interconnect width and spacing each are equal to 0.54 μm. The comparison of output voltage waveforms generated by SPICE simulations and analytical model for in-phase switching under fast and slow ramps has been shown in Figs. 4.5 and 4.6, respectively.

It can be observed that the proposed model waveform matches closely with SPICE waveform. For fast ramp input, NMOS width (W_{n1}) of 97.5 nm has been taken, while for slow ramp, W_{n1} = 4 μm has been taken. Furthermore, for the CMOS drivers, PMOS channel width is 2.5 times the NMOS width. The values of parasitic impedance parameters for the two lines are extracted from [222]. These line parasitics are the following: line resistance, $R_1 = R_2 = 208.93\ \Omega$ and line capacitance, $C_1 = C_2 = 301.475$ fF. The two lines are coupled through the coupling capacitance of 105.4 fF.

For fast ramp input, average propagation delay is analytically determined for in-phase switching and has been provided in Table 4.1 along with computational error wrt SPICE simulation results. Varying interconnect load conditions and

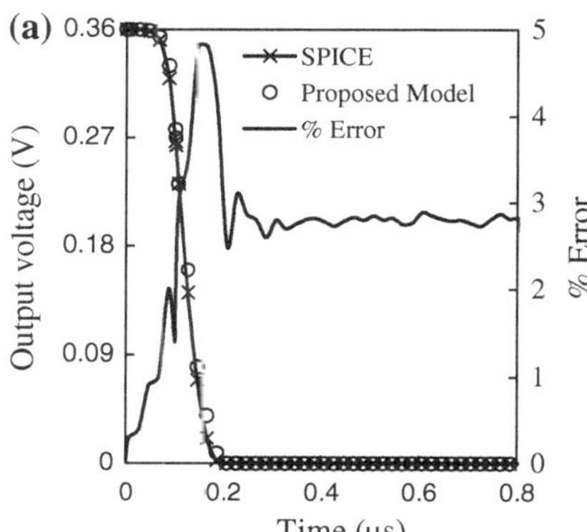

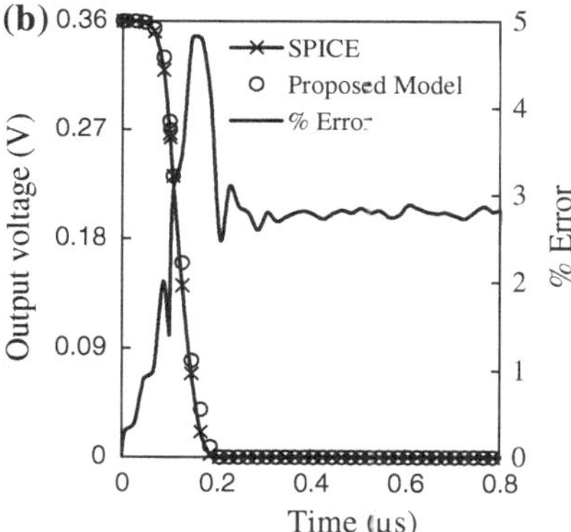

Fig. 4.5 Voltage waveform at the output of **a** aggressor driver, **b** victim driver under in-phase switching for fast ramp with $W_{\text{n1}} = 97.5\ \text{nm} = W_{\text{n2}}$, $W_{\text{p1}} = 2.5W_{\text{n1}} = W_{\text{p2}}$

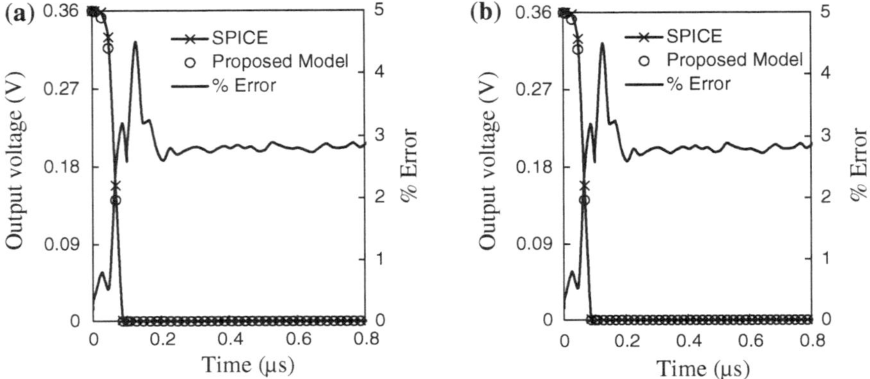

Fig. 4.6 Voltage waveform at the output of **a** aggressor driver, **b** victim driver under in-phase switching for slow ramp with $W_{n1} = 4$ μm $= W_{n2}$, $W_{p1} = 2.5W_{n1} = W_{p2}$

widths for the aggressor and victim drivers have been considered. The proposed analytical model yields maximum errors in the propagation delays for the aggressor and victim drivers as 6.99 and 2.75 %, respectively, whereas the average errors involved in the same are 3.34 and 1.71 %, respectively.

Figures 4.7 and 4.8 confirm the validity of the proposed model by comparing the waveforms generated analytically and SPICE simulations under out-of-phase switching. PMOS victim widths (W_{p2}) of 0.4 and 10 μm have been considered for fast and slow ramp inputs, respectively. It can be observed that analytical results match SPICE simulations quite closely.

Table 4.2 presents an account of the propagation delay and computational error involved as predicted by the proposed model wrt SPICE simulations for out-of-phase switching. Variable interconnect load and asymmetric aggressor and victim driver dimensions have been considered.

It is observed that τ_{p1} obtained by the proposed model has an average error of 3.44 % and maximum error of 6.15 %. Similarly, τ_{p2} predicted by the proposed analytical model results in average and maximum errors of 2 and 3.91 %, respectively. It is to be noted that delay estimates provided in Tables 4.1 and 4.2 are based on fast ramp input.

Table 4.3 compares the timing of aggressor and victim drivers wrt SPICE simulation. Here, timing refers to the instant when active transistors make transition from sub-saturation to sub-linear region of operation. Fast and slow ramps have been considered for in-phase and out-of-phase transitions. It can be seen that maximum errors in the estimation of timing for the aggressor and victim drivers under in-phase switching are 8.120 and 7.445 %, while the average errors in the same are 5.197 and 4.571 %. For out-of-phase switching, the maximum and average errors predicted by the proposed analytical model are 9.827, 7.662 % and 4.257, 4.613 % for the aggressor and victim drivers, respectively. Thus, transition time is very well predicted by the proposed model.

Table 4.1 Propagation delay and error involved for in-phase switching wrt SPICE simulation

W_{n1} (nm)	W_{n2} (nm)	Circuit parameters					Propagation delay				Error in τ_{p1} (%)	Error in τ_{p2} (%)
							SPICE		Analytical			
		R_1 (Ω)	R_2 (Ω)	C_1 (pf)	C_2 (pf)	C_c (pf)	τ_{p1} (μs)	τ_{p2} (μs)	τ_{p1} (μs)	τ_{p2} (μs)		
97.5	97.5	208.93	208.93	0.30	0.30	0.11	0.1183	0.1183	0.1151	0.1151	2.67	2.67
97.5	97.5	208.93	208.93	0.20	0.30	0.09	0.1090	0.1162	0.1080	0.1130	0.94	2.75
97.5	160	208.93	208.93	0.50	0.20	0.11	0.1292	0.1014	0.1202	0.1017	6.99	0.26
160	97.5	208.93	208.93	0.50	0.20	0.11	0.1159	0.1092	0.1135	0.1073	2.07	1.77
97.5	292.5	208.93	208.93	0.30	0.30	0.20	0.1071	0.0969	0.1028	0.0958	4.03	1.12

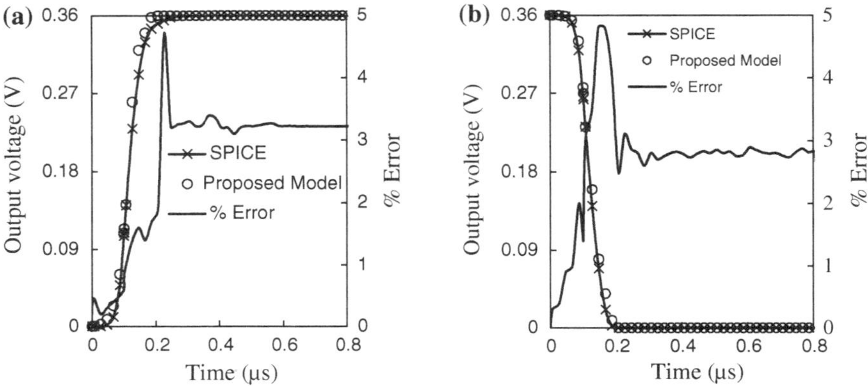

Fig. 4.7 Voltage waveform at the output of **a** aggressor driver, **b** victim driver under out-of-phase switching for fast ramp with W_{n1} = 0.10 μm, W_{n2} = 0.16 μm, W_{p1} = 0.24 μm, W_{p2} = 0.4 μm

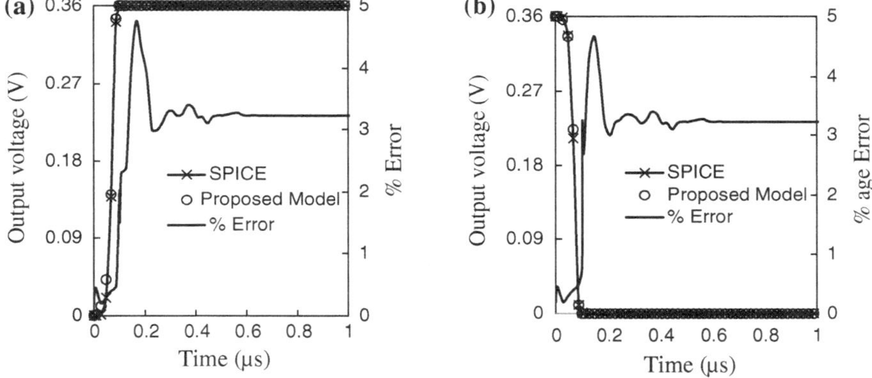

Fig. 4.8 Voltage waveform at the output of **a** aggressor driver, **b** victim driver under out-of-phase switching for slow ramp with W_{n1} = 4 μm = W_{n2}, W_{p1} = 10 μm = W_{p2}

For in-phase and out-of-phase transitions, variation of propagation delay with the aggressor and victim MOS dimensions is shown in Table 4.4. The error involved in propagation delay wrt SPICE simulations under switching conditions considered is also computed. Aggressor MOS width (W_{n1}) is varied from 0.1 to 4 μm. The propagation delay of aggressor and victim drivers predicted by the proposed model for in-phase switching exhibit maximum errors of 2.77 and 2.70 %, respectively, wrt SPICE. Under out-of-phase switching, the propagation delay estimated by the proposed model has maximum errors of 8.04 and 5.65 % for MN1 and MP2, respectively. It can also be observed from Table 4.4 that as aggressor width is increased from 0.1 to 4 μm, propagation delay decreases by 45.6 and 50.1 % for in-phase and out-of-phase transitions, respectively.

Table 4.2 Propagation delay and error involved for out-of-phase switching wrt SPICE simulation

W_{n1} (nm)	W_{n2} (nm)	Circuit parameters					Propagation delay				Error in τ_{p1} (%)	Error in τ_{p2} (%)
							SPICE		Analytical			
		R_1 (Ω)	R_2 (Ω)	C_1 (pf)	C_2 (pf)	C_c (pf)	τ_{p1} (μs)	τ_{p2} (μs)	τ_{p1} (μs)	τ_{p2} (μs)		
97.5	243.7	208.93	208.93	0.3	0.3	0.11	0.1406	0.1408	0.1362	0.1440	3.13	2.28
97.5	243.7	208.93	208.93	0.3	0.6	0.11	0.1362	0.1798	0.1329	0.1780	2.44	1.00
97.5	243.7	208.93	208.93	0.6	0.3	0.11	0.1789	0.1361	0.1700	0.1360	4.96	0.07
97.5	400	208.93	208.93	0.6	0.9	0.11	0.1754	0.1563	0.1646	0.1540	6.15	1.48
97.5	400	208.93	208.93	0.6	0.6	0.11	0.1792	0.1352	0.1710	0.1340	4.57	0.86
160	243.7	208.93	208.93	0.6	0.6	0.11	0.2203	0.2662	0.2204	0.2570	0.03	3.47
160	400	208.93	208.93	0.6	1.2	0.21	0.2307	0.2885	0.2310	0.2800	0.11	2.94
160	400	208.93	208.93	1.2	0.60	0.32	0.2961	0.2435	0.2780	0.2340	6.13	3.91

Table 4.3 Aggressor and victim timings along with the error involved in timing wrt SPICE simulation

Switching type	Ramp type	Aggressor MOS width (μm)	Aggressor timing (μs)		Error (%)	Victim MOS width (μm)	Victim timing (μs)		Error (%)
			SPICE	Analytic			SPICE	Analytic	
In-phase	Fast	0.1	0.137	0.1268	7.445	0.1	0.137	0.1268	7.445
		0.1	0.133	0.1222	8.120	0.16	0.119	0.1134	4.706
	Slow	2	0.0757	0.0743	1.876	4	0.717	0.0697	2.789
		4	0.0717	0.0693	3.347	4	0.0717	0.0693	3.347
Out-of-phase	Fast	0.1	0.173	0.1560	9.827	0.24	0.177	0.1734	2.034
		0.1	0.177	0.1703	3.785	0.4	0.135	0.131	2.963
	Slow	2	0.0837	0.0849	1.434	10	0.0757	0.0815	7.662
		4	0.0757	0.0772	1.982	10	0.0777	0.0732	5.792

Table 4.4 Propagation delay with aggressor and victim driver size and error involved wrt SPICE simulation

Switching type	Ramp type	Aggressor MOS width (μm)	Propagation delay (μs)		Error (%)	Victim MOS width (μm)	Propagation delay (μs)		Error (%)
			SPICE	Analytic			SPICE	Analytic	
In-phase	Fast	0.1	0.1183	0.1151	2.70	0.1	0.1183	0.1151	2.70
		0.1	0.1149	0.1118	2.77	0.16	0.1059	0.1056	0.28
	Slow	2	0.0705	0.0698	1.10	4	0.0654	0.0653	0.11
		4	0.0643	0.0650	1.04	4	0.0643	0.0650	1.04
Out-of-phase	Fast	0.1	0.1406	0.1363	3.06	0.24	0.1408	0.1438	2.14
		0.1	0.1469	0.1468	0.08	0.4	0.1157	0.1150	0.61
	Slow	2	0.0759	0.082	8.04	10	0.0702	0.0726	3.45
		4	0.0701	0.0733	4.64	10	0.0710	0.0670	5.65

Another observation of the analysis is that propagation delay is higher under out-of-phase switching. It is observed that when drivers are switching in the same direction, the delay is decreased, known as positive effect. When transitions are opposite, the delay is increased, known as negative effect. This is accounted for by the fact that amount of interconnect coupling capacitance is dependent upon the nature of the signal transitions [92–94]. If drivers are driven by signals switching in the same direction, the effective coupling capacitance is approximately zero and the total capacitance of each interconnect is approximately equal to the line-to-ground capacitance. Alternatively, if signals on each interconnect are switching in the opposite direction or out-of-phase, the effective capacitance approximately doubles to $2 \times C_c$. Hence, the delay variations can be positive or negative, depending on the direction of the simultaneous transitions. Subsequently, if the signals on both interconnects are in-phase, the total capacitance is lower, reducing the average power dissipation. Alternatively, out-of-phase signals increase the total capacitance, thereby increasing the overall power dissipation.

A similar trend is observed for power-delay-product which is characterized as the quality metric under simultaneously switching coupled scenario. The variations of power-delay-product with aggressor driver width are shown in Fig. 4.9a. It can be seen that PDP increases with increase in aggressor driver width. This is because

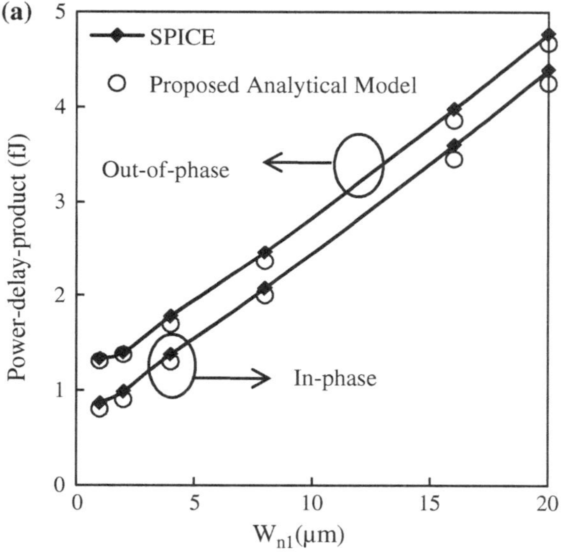

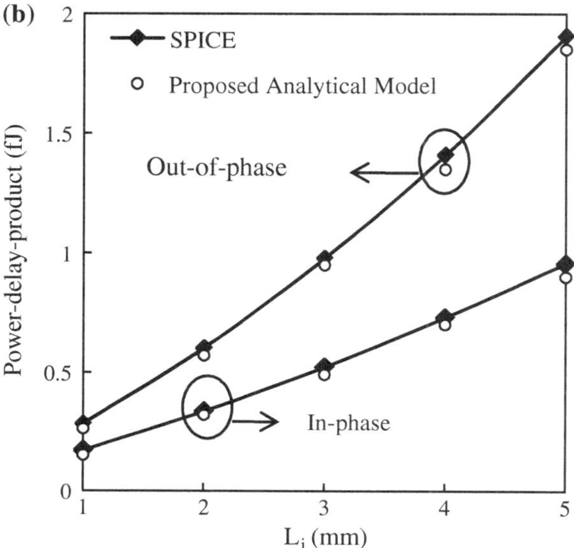

Fig. 4.9 SPICE and analytic variations of power–delay –product under in-phase and out-of-phase switching for 65-nm technology with **a** aggressor driver width and **b** interconnect length

the decrease in the propagation delay more than offsets increase in power dissipation due to the aggressor upsizing. The close proximity of the proposed analytical model with SPICE is also observed. Drivers in the two-line coupled system are equal and balanced with $W_{n1} = W_{n2} = 0.1$ μm and $W_{p1} = 2.5W_{n1} = W_{p2}$ in the respective drivers. The input signal rise time is taken as 0.1 μs, and interconnect length is 5 mm.

The variation of PDP with the interconnect length is shown in Fig. 4.9b. It can be observed that PDP increases with increase in the interconnect length and is considerably higher for longer interconnect lengths. This is because both power dissipation and delay increase with interconnect length. Furthermore, PDP increases sharply for out-of-phase transition in sharp contrast to in-phase switching. This can be attributed to the Miller's effect where coupling capacitance is effectively doubled under out-of-phase switching. Due to this, the charging/discharging of interconnect lines is quite slow, causing overall delay to increase considerably. For example, SPICE-extracted power-delay-product increases from 0.17 to 0.95 fJ for in-phase switching case. However, PDP increases from 0.28 to 1.90 fJ under out-of-phase switching case. Similar observations are observed for 130-nm and 90-nm technology nodes as well.

4.5 Concluding Remarks

In this chapter, crosstalk analysis of capacitively coupled CMOS buffer-driven interconnects for in-phase and out-of-phase switching conditions is presented. The analysis has been carried out for 65-nm technology node. A comparative analysis between the proposed analytical model and SPICE results shows a good agreement. Comparison of the proposed analytical models with SPICE shows that the analytical results capture waveform shape, propagation delay, and timings with good accuracy. Varying load conditions have been considered.

Under in-phase switching, the average error in the propagation delay with respect to SPICE is 3.34 and 1.71 % for the aggressor and victim drivers, respectively. For out-of-phase switching, average errors in the same are 3.44 and 2 %. The timing is also very well predicted by the proposed models. The average errors involved in the estimation of timing for the aggressor and victim drivers are 5.20 and 4.57 %, under in-phase switching. The average errors involved in the same for out-of-phase switching are 4.26 and 4.61 %. The close agreement between these results and those obtained by SPICE simulations clearly shows that the proposed approach shall be very useful for ultra-low-power electronic design. Since power-delay-product increases with the interconnect length and aggressor driver width, it is advantageous to keep buffer dimensions and interconnect length smaller. The present chapter thus provides comprehensive analysis of dynamic crosstalk in subthreshold regime. The functional crosstalk analysis when aggressor driver is switching and the victim driver quiet has been dealt with in the next chapter.

Chapter 5
Subthreshold Interconnect Noise Analysis

Keywords Aggressor/victim driver · Coupling noise · Integrated circuit · SPICE · Step input

Interconnects in a CMOS integrated circuit are the conductors deposited on the dielectric. In integrated circuits, interconnect lines are driven by CMOS logic gates. For either an in-phase or out-of-phase transition, the coupling capacitance affects the waveform shape of the output voltage and the propagation delay of each inverter, primarily changing the speed of a CMOS integrated circuit. If one of these CMOS logic gates is quiet, while other logic gates are in transition, the coupling capacitance induces overshoots (signal rises above the ground) and undershoots (signal falls below the ground) on the victim (quiet) line because of switching activity on the aggressor (active) line. This unwanted interference from neighboring signal wire to a network node is referred to as functional crosstalk or coupling noise and seriously affects the circuit behavior, dissipates unnecessary dynamic power, and introduces delay uncertainty within the circuit. Estimating the peak and timing of this coupling noise is therefore important in order to ensure signal integrity and avoid malfunctions.

One major advantage of CMOS digital circuits is that CMOS logic gates are relatively immune to noisy environment [13]. In subthreshold, however, the ratio of supply voltage to the transistor threshold voltage is less than unity, and this advantage significantly gets diminished. Therefore, the problem of noise exaggerates in importance such that coupling noise becomes a serious threat to the continued growth in integration density and subthreshold circuit performance [9]. In addition to the voltage change at the output of victim node, crosstalk also increases the delay of active logic gates. The increased delay is not a concern for subthreshold circuits since these tend to be low performance. However, there is very less work exploring the effect of device subthreshold operation on coupling noise from the perspective of modeling approach. Mathematical models in this direction if developed will be very useful, as it is an essential area of research for ultra-low-power applications.

Consequently, in this chapter analytical expressions for the output voltage and coupling noise voltage in subthreshold regime for CMOS buffer-driven coupled resistive–capacitive VLSI interconnects have been developed. Based on these

R. Dhiman and R. Chandel, *Compact Models and Performance Investigations for Subthreshold Interconnects*, Energy Systems in Electrical Engineering,
DOI 10.1007/978-81-322-2132-6_5

expressions, the propagation delay and peak coupling noise voltage are determined. The analysis has been carried out for 65-nm technology node. The accuracy of analytical expressions is compared with SPICE.

5.1 *Inv*1 Input Switching from Low-to-High and *Inv*2 Static High

In the following analysis, it is assumed that inverter 1 (*Inv*1) input shown in Fig. 4.2 transits from low-to-high, while input of inverter 2 (*Inv*2) is quiet (static high). Thus, *Inv*1 is considered as aggressor and *Inv*2 victim. Under such input condition, the coupling capacitance induces undershoots at the output of quiet inverter. This coupled noise voltage may result in the logic failure of the VLSI circuit.

A simplified circuit model to analyze the coupling noise voltage at the output of quiet inverter and propagation delay of the active inverter is shown in Fig. 5.1. The related current directions are also shown. Current through PMOS transistor has been neglected under the assumption of fast ramp input.

The signal at the input of *Inv*1 is assumed to be shaped as a rising ramp signal and is given by

$$V_{\text{in1}} = V_{\text{DD}} \frac{t}{\tau_{\text{r}}} \quad 0 \leq t \leq \tau_{\text{r}} \tag{5.1}$$

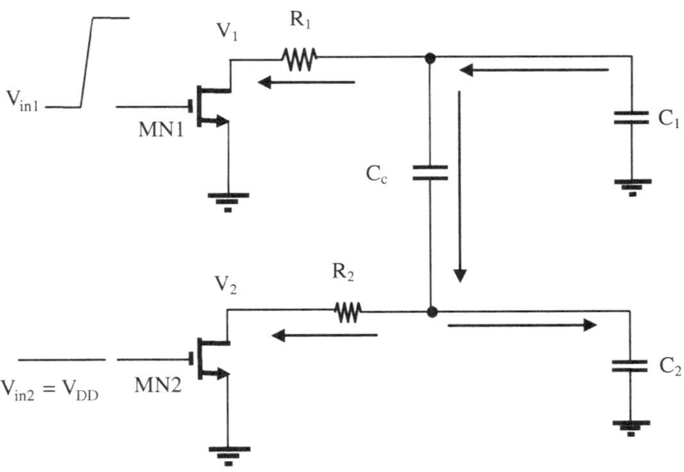

Fig. 5.1 Buffer-driven interconnects for aggressor input switching from low-to-high and victim input at static high

Input signal driving *Inv*2 is equal to V_{DD}, i.e.,

$$V_{in2} = V_{DD} \tag{5.2}$$

The current through MP1 transistor is small and is neglected under the assumption of fast ramp. The differential equations governing the output voltage of each MOS transistor are as given by

$$-(C_1 + C_c)\frac{dV_1}{dt} + C_c\frac{dV_2}{dt} = I_{n1} + R_1(C_1 + C_c)\frac{dI_{n1}}{dt} - R_2C_c\frac{dI_{n2}}{dt} \tag{5.3}$$

$$C_c\frac{dV_1}{dt} - (C_2 + C_c)\frac{dV_2}{dt} = I_{n2} - R_1C_c\frac{dI_{n1}}{dt} + R_2(C_2 + C_c)\frac{dI_{n2}}{dt} \tag{5.4}$$

There are no tractable solutions to the differential equations (5.3) and (5.4). In order to obtain the output voltages V_1 and V_2 of *Inv*1 and *Inv*2, respectively, it is necessary to make certain simplifying assumptions.

5.1.1 Step Input Approximation

The input can be approximated as a step input if the transition time of the input signal is smaller compared to the propagation delay of CMOS inverters and the output transition time [96]. Under the step input approximation, differential equations, (5.3) and (5.4), become

$$-(C_1 + C_c)\frac{dV_1}{dt} + C_c(1 + \gamma_{n2}R_2)\frac{dV_2}{dt} = B_{n1} \tag{5.5}$$

$$C_c\frac{dV_1}{dt} - (C_2 + C_c)(1 + \gamma_{n2}R_2)\frac{dV_2}{dt} = \gamma_{n2}V_2 \tag{5.6}$$

Analytical expressions characterizing the output voltage of *Inv*1 and the coupling noise voltage at the output of *Inv*2 before MN1 starts to operate in the linear region are

$$V_1 = V_{DD} - \frac{B_{n1}}{C_1 + C_c}t + \frac{C_c}{C_1 + C_c}(1 + \gamma_{n2}R_2)V_2 \tag{5.7}$$

$$V_2 = -\frac{B_{n1}C_c}{\gamma_{n2}(C_1 + C_c)}(1 - e^{-\alpha_{n2}t}) \tag{5.8}$$

The propagation delay of *Inv*1 ($t_{P_{HL1}}$) is computed using (5.7) and Newton–Raphson iteration as

$$\frac{B_{n1}}{C_1 + C_c} t_{P_{HL1}} + \frac{C_c}{C_1 + C_c}(1 + \gamma_{n2} R_2) V_2 = 0.5 V_{DD} \tag{5.9}$$

Since the current through MN2 discharges the load capacitance C_1, the propagation delay is less than the estimated delay based on the load of $C_1 + C_c$.

The peak of undershoot voltage occurs at τ_{nsat}^1 and is determined as

$$V_2(\text{peak}) = -\frac{B_{n1} C_c}{\gamma_{n2}(C_1 + C_c)}\left(1 - e^{-\alpha_{n2}\tau_{nsat}^1}\right) \tag{5.10}$$

It can be seen from (5.10) that peak coupling noise voltage is proportional to B_{n1}/γ_{n2} and C_c. Thus, if the effective output conductance of MN2 is increased, the peak noise voltage can be reduced. This conclusion suggests that the size of the MOS transistors within the quiet inverter should be increased. It can also be noted that peak coupling noise voltage is also proportional to B_{n1}, therefore increasing the aggressor driver size increases the peak noise. Furthermore, V_2 decreases exponentially in the sub-linear region.

The duration time τ_{nsat}^1 when MN1 leaves the sub-saturation region is determined from (5.7) using the condition that at $t = \tau_{nsat}^1$, $V_1 = 4U_{th}$, for this data, Eq. (5.7) provides

$$V_{DD} - \frac{B_{n1}}{C_1 + C_c}\tau_{nsat}^1 + \frac{C_c}{C_1 + C_c}(1 + \gamma_{n2} R_2) V_2 = 4U_{th} \tag{5.11}$$

From (5.11), τ_{nsat}^1 is obtained by using Newton–Raphson method. After τ_{nsat}^1, both active transistors operate in the linear region. The analytical solutions for this region have been obtained in Sect. 4.2.

5.1.2 Neglecting Current Through MN2

The analysis described in this subsection is based on the assumption that current through MN2 is negligible, i.e., $\gamma_{n2} V_2$ is negligible compared to $C_c dV_1/dt$. Based on this assumption, the differential equations governing the output voltages are given by

$$-(C_1 + C_c)\frac{dV_1}{dt} + C_c\frac{dV_2}{dt} = I_{n1} + R_1(C_1 + C_c)\frac{dI_{n1}}{dt} \tag{5.12}$$

$$C_c\frac{dV_1}{dt} - (C_2 + C_c)\frac{dV_2}{dt} = -R_1 C_c\frac{dI_{n1}}{dt} \tag{5.13}$$

The solutions of V_1 and V_2 in the range $0 \le t \le \tau_r$ are given as

$$V_1 = V_{DD} - R_1 B_{n1} e^{-\frac{V_{DD}}{\eta_n U_{th}}} \left[e^{\frac{V_{DD}}{\tau_r \eta_n U_{th}} t} - 1 \right] - \frac{C_2 + C_c}{C_1 C_2 + C_c(C_1 + C_2)} V_{n,1} \tag{5.14}$$

$$V_2 = -\frac{C_c}{C_1 C_2 + C_c(C_1 + C_2)} \frac{B_{n1} \tau_r \eta_n U_{th}}{V_{DD}} e^{-\frac{V_{DD}}{\eta_n U_{th}}} \left[e^{\frac{V_{DD}}{\tau_r \eta_n U_{th}} t} - 1 \right] \tag{5.15}$$

When the input signal reaches V_{DD} at τ_r, MN1 continues to operate in the sub-saturation region. The coupling noise voltage at the end of input transition is

$$V_2(\tau_r) = -\frac{C_c}{C_1 C_2 + C_c(C_1 + C_2)} \frac{B_{n1} \tau_r \eta_n U_{th}}{V_{DD}} \left[1 - e^{-\frac{V_{DD}}{\eta_n U_{th}}} \right] \tag{5.16}$$

After the input transition is completed, the current through MN2 cannot be neglected since $\gamma_{n2} V_2$ may be comparable to $C_c dV_1/dt$. After τ_r, the output voltages V_1 and V_2 before MN1 enters into the sub-saturation region are described by

$$\begin{aligned} V_1 = {} & V_1(\tau_r) - \frac{B_{n1}}{C_1 + C_c}(t - \tau_r) \\ & - \frac{C_c}{C_1 + C_c}(1 + \gamma_{n2} R_2)\left(V_2(\tau_r) + V_{2,a}\right)\left(1 - e^{-\alpha_{n2}(t - \tau_r)}\right) \end{aligned} \tag{5.17}$$

$$V_2 = -V_{2,a} + \left[V_2(\tau_r) + V_{2,a} \right] e^{-\alpha_{n2}(t - \tau_r)} \tag{5.18}$$

where

$$\alpha_{n2} = \frac{\gamma_{n2}}{1 + \gamma_{n2} R_2} \frac{C_1 + C_c}{C_1 C_2 + C_c(C_1 + C_2)} \tag{5.19}$$

$$V_{2,a} = \frac{C_c}{(C_1 + C_c)\gamma_{n2}} B_{n1} \tag{5.20}$$

The time when MN1 leaves the sub-saturation region and the propagation delay time are determined from (5.11) by applying Newton–Raphson iteration. The peak coupling noise voltage is approximated to occur at τ^1_{nsat} and is equal to $V_2\left(\tau^1_{nsat}\right)$ which is determined from (5.12).

5.2 *Inv*1 Input Switching from Low-to-High and *Inv*2 Static Low

In this case, it is assumed that the input of *Inv*2 is at ground, while *Inv*1 transitions from high-to-low. This is shown in Fig. 5.2. MN1 and MP2 are the active transistors in each inverter. The initial values of both V_1 and V_2 are V_{DD}.

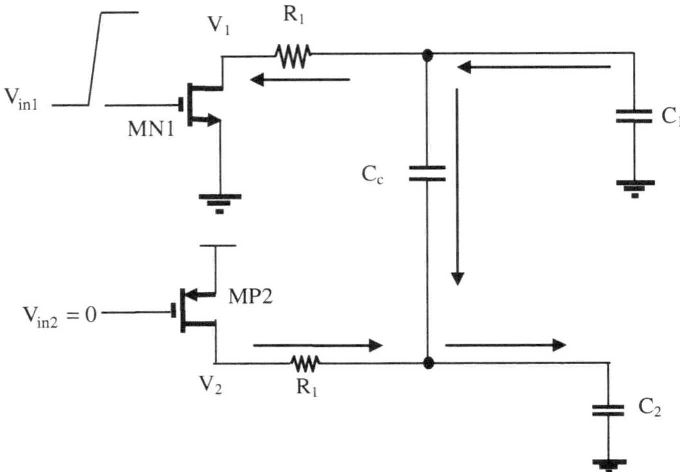

Fig. 5.2 Buffer-driven interconnects for aggressor input switching from low-to-high and victim at static low

Owing to the coupling capacitance, overshoots are exhibited at the output of quiet inverter when aggressor driver switches from high-to-low.

The input to *Inv*1 is a ramp signal which is given by (5.1). The differential equations governing the output voltage of *Inv*1 and *Inv*2 shown in Fig. 5.2 are given by

$$(C_1 + C_c)\frac{dV_1}{dt} - C_c\frac{dV_2}{dt} = I_{p1} + R_1(C_1 + C_c)\frac{dI_{p1}}{dt} + R_2C_c\frac{dI_{n2}}{dt} \tag{5.21}$$

$$C_c\frac{dV_1}{dt} - (C_2 + C_c)\frac{dV_2}{dt} = I_{n2} + R_1C_c\frac{dI_{p1}}{dt} + R_2(C_2 + C_c)\frac{dI_{n2}}{dt} \tag{5.22}$$

Here, I_{p1} and I_{n2} are the currents that flow across MP1 and MN2, respectively. Similar to the previous analysis, there are no tractable solutions to the differential equations (5.21) and (5.22). Analytical expressions characterizing the output voltages V_1 and V_2 have been derived under simplifying assumption of step input.

5.2.1 Step Input Approximation

The differential equations (5.21) and (5.22) under the assumption of step input reduce to

$$(C_1 + C_c)\frac{dV_1}{dt} - C_c\frac{dV_2}{dt} = I_{p1} + R_2C_c\frac{dI_{n2}}{dt} \tag{5.23}$$

$$C_c \frac{dV_1}{dt} - (C_2 + C_c)\frac{dV_2}{dt} = I_{n2} + R_2(C_2 + C_c)\frac{dI_{n2}}{dt} \tag{5.24}$$

Analytical expressions characterizing the output voltage of *Inv*1 and the coupling noise voltage at the output of *Inv2* before MN1 starts to operate in the linear region are given by

$$V_1 = V_{DD} - \frac{B_{n1}}{C_1 + C_c}t + \frac{C_c}{C_1 + C_c}V_{p2} \tag{5.25}$$

$$V_2 = V_{DD} - \frac{C_c B_{n1}}{(C_1 + C_c)\gamma_{p2}}(1 - e^{-\alpha_{p2}t}) \tag{5.26}$$

where

$$V_{p2} = \frac{C_c}{C_1 + C_c}B_{n1}(1 - e^{-\alpha_{p2}t}) \tag{5.27}$$

$$\alpha_{p2} = \frac{\gamma_{p2}}{1 + \gamma_{p2}R_2}\frac{C_1 + C_c}{C_1C_2 + C_c(C_1 + C_2)} \tag{5.28}$$

From (5.26), it is found that the coupling noise voltage is proportional to B_{n1}/γ_{p2} and C_c. Thus, if the output conductance of MP2 is increased and the coupling capacitance is decreased, the peak of overshoots gets reduced.

The peak overshoot across *Inv*2 is approximated to occur at $t = \tau^1_{nsat}$ and is given by

$$V_2(\text{peak}) = V_{DD} - \frac{C_c B_{n1}}{(C_1 + C_c)\gamma_{p2}}\left(1 - e^{-\alpha_{p2}\tau^1_{nsat}}\right) \tag{5.29}$$

The time duration τ^1_{nsat} when MN1 leaves the sub-saturation region and propagation delay time $t_{P_{HL1}}$ are determined using (5.25) and Newton–Raphson iteration. Since the current through MP2 slows down the discharge process, the propagation delay is greater than the delay estimated assuming $C_1 + C_c$ as the load capacitance. After τ^1_{nsat}, both active transistors operate in the sub-linear region. The analytical solutions for this region have been obtained in Sect. 4.2.

The output voltage waveforms for the case when *Inv*1 switches from high-to-low and input to the victim driver is kept high are shown in Fig. 5.3a, b, using the proposed analytical approach and those obtained by SPICE simulations. Interconnect length of 5 mm has been taken. A good agreement between the proposed model and SPICE results is obtained. The error in the propagation delay of the aggressor driver using the analytical model is 6.28 % compared to SPICE. The voltage waveform of victim driver is shown in Fig. 5.3b. The peak negative noise voltage predicted by the analytical model is 11.3 mV with 1.73 % error with respect to SPICE.

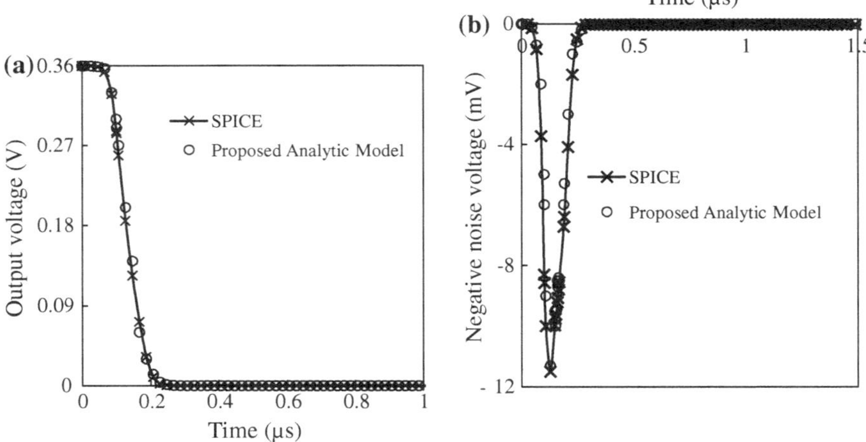

Fig. 5.3 Output voltage waveforms for **a** aggressor driver, **b** victim driver: $W_{n1} = W_{n2} = 97.5$ nm, $W_{p1} = W_{p2} = 2.5W_{n1}$, $C_1 = C_2 = 301.475$ fF, $C_c = 105.395$ fF, $R_1 = R_2 = 208.26\ \Omega$

The verification of the proposed analytic approach and SPICE for the case when aggressor driver switches from high-to-low and input of *Inv*2 is kept at static low has been shown in Figs. 5.4 and 5.5. Two further test cases are considered. The test cases consider different configurations of aggressor and victim lines and are as given below:

(i) $L_i = 3$ mm, $s = 8$ μm, $w = 0.48$ μm, R_1, $R_2 = 140.4\ \Omega$, C_1, $C_2 = 237.08$ fF, $C_c = 13.803$ fF
(ii) $L_i = 4$ mm, $s = 2$ μm, $w = 0.6$ μm, R_1, $R_2 = 150.43\ \Omega$, C_1, $C_2 = 247.82$ fF, $C_c = 85.444$ fF

where w and s refer, respectively, to the line width and interconnect spacing.

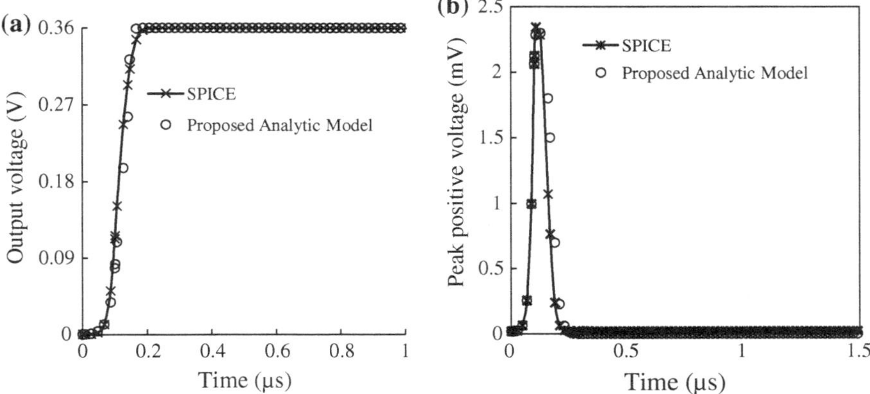

Fig. 5.4 Output voltage waveforms for **a** aggressor driver, **b** victim driver for test case (i)

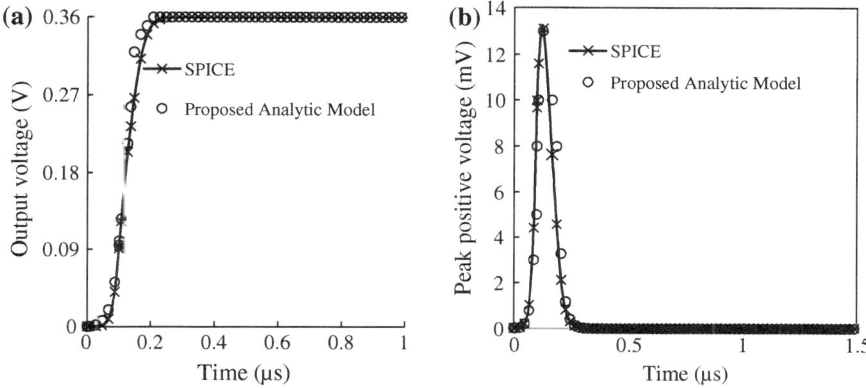

Fig. 5.5 Output voltage waveforms for **a** aggressor driver, **b** victim driver for test case (ii)

Interconnect impedance parasitics viz R_1, R_2, C_1, C_2, and C_c are obtained using the PTM. The percentage errors in the propagation delays of *Inv*1 as predicted by the proposed models with respect to SPICE are 9.12 and 5.8 %, respectively, for the two test cases under consideration. The percentage errors in the estimation of peak noise voltage are 0.76 and 2.6 % in Figs. 5.4b and 5.5b, respectively. Thus, proposed analytical models capture propagation delay and peak noise voltage in the closer proximity of SPICE simulations.

5.3 Design Guidelines for Crosstalk Avoidance

In this analysis, aggressor line width and line-to-line spacing are varied. The dependency of negative and positive noise voltages on interconnect geometric parameters (w, s) is shown, respectively, in Figs. 5.6 and 5.7. Increasing the width of aggressor line increases the line-to-ground capacitance and reduces the parasitic resistance. An increase in the ground capacitance decreases the coupling noise due to the lower impedance offered by the line-to-ground capacitance. The increased ground capacitance behaves as a filter, thereby reducing the coupling noise. Interconnect impedance parasitics for different line width and spacing have been provided in Table 5.1.

It can also be observed that increased spacing between the wires reduces the peak noise voltage. It is because increased spacing decreases the coupling capacitance. For example, in Fig. 5.6a, the peak noise voltage is −27 mV for aggressor line width and spacing each equal to 0.54 μm. The peak negative noise reduces to 18.9, 7.19, and 2.35 mV as w and s, respectively, are varied in the following order: (0.8, 1 μm), (1, 3 μm), and (2, 8 μm).

Similarly, in Fig. 5.7b, peak positive noise decreases from 93.1 to 40.7 mV, 10.8 and 4.54 mV as w and s, respectively, are varied in the order (0.54, 0.54 μm)

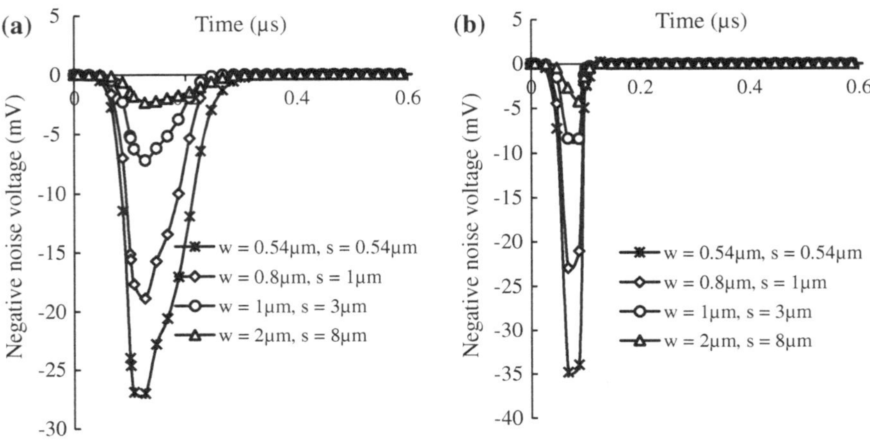

Fig. 5.6 Negative noise voltage waveforms: **a** L_i = 2 mm, $W_{n1} = W_{n2}$ = 97.5 nm, **b** L_i = 2 mm, W_{n1} = 2 μm, W_{n2} = 97.5 nm

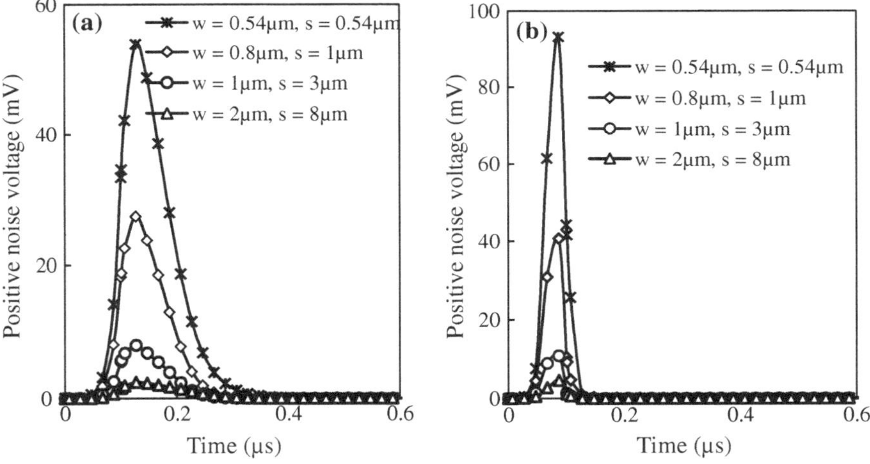

Fig. 5.7 Positive noise voltage waveforms: **a** L_i = 4 mm, W_{p1} = 243.7 nm, W_{n2} = 97.5 nm, **b** L_i = 2 mm, W_{p1} = 5 μm, W_{n2} = 97.5 nm

Table 5.1 Interconnect parasitics for various configurations

s, w (μm)	R_1, R_2 (Ω)	C_1, C_2 (fF)	C_c (fF)
0.54, 0.54	83.57	68.246	153.504
0.8, 1	56.41	106.108	88.62
1, 3	45.128	161.108	30.242
2, 8	22.564	242.056	12.51

(0.8, 1 μm), (1, 3 μm), and (2, 8 μm). Thus, increasing the aggressor line width and spacing are effective noise avoidance techniques.

An important observation is that with increase in the aggressor driver width (W_{n1}), peak crosstalk increases. For example, in Fig. 5.6, peak negative noise voltage increases from 27 to 34.8 mV as W_{n1} is increased from 97.5 nm to 2 μm. Interconnect width and spacing both are equal to 0.54 μm. This conclusion confirms the fact that peak crosstalk increases with the aggressor driver width (Eq. 5.10).

The relative sizes of aggressor and victim drivers also affect the coupling noise, and coupling noise induced delay variation [227, 228]. The effect of victim driver width on peak negative noise voltage is shown in Fig. 5.8. Line-to-line spacing is varied from 0.5 to 6 μm.

It can be observed that increasing victim driver width reduces the peak negative noise. This is because increasing victim driver size increases its output conductance. This reduces the noise voltage that the victim driver can hold since the victim is more effectively connected to a stable voltage, i.e., ground. For instance, peak negative noise reduces by 95.2 % when MN2 is made 20 times larger, s being equal to 0.5 μm. Victim driver sizing is thus an effective noise avoidance technique. Increasing the size of the driver on the victim, however, increases the overall area, making this technique subject to area constraints. It is also seen that peak negative noise saturates for larger victim driver widths. This is owing to the fact that gate capacitance of the victim driver increases which compensates for any reduction in the peak negative noise caused by larger driver widths. Thus, only a limited increase in victim driver sizing is advantageous. Alternatively, downsizing aggressor driver decreases the coupling noise since the ability of the aggressor to induce noise on the victim line is reduced. Decreasing the size of the driver on the aggressor, however, reduces the coupling noise at the expense of increased delay. Adjusting size of the aggressor driver to reduce crosstalk is therefore subject to delay constraints. Furthermore, peak negative noise decreases by 2.34 mV

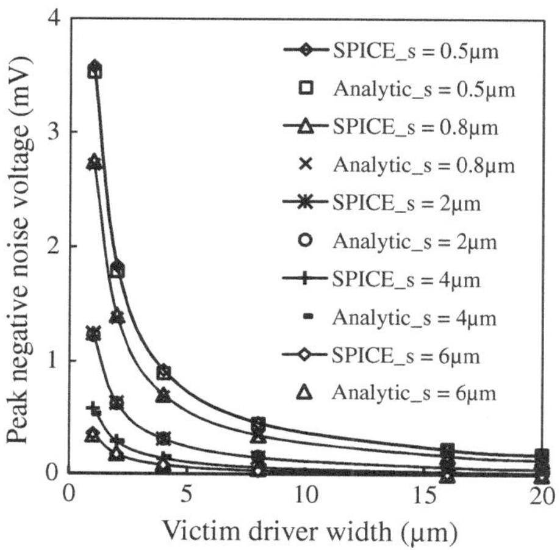

Fig. 5.8 Peak negative noise voltage with victim driver width for different line-to-line spacing

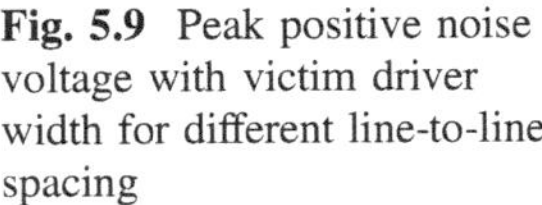

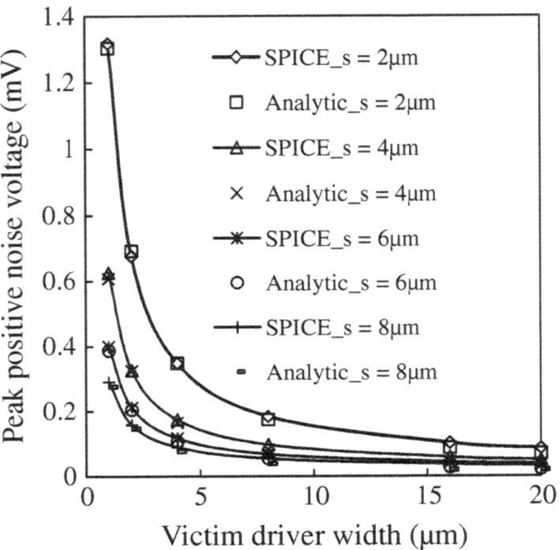

Fig. 5.9 Peak positive noise voltage with victim driver width for different line-to-line spacing

(65 % reduction) when s is varied from 0.5 to 2 μm. Similar observations are also made for peak positive noise voltage as shown in Fig. 5.9.

In this case also, analytical results obtained using the proposed models track SPICE simulation results very closely. Peak positive noise voltage reduces by 0.69 mV (52 %) when spacing is increased from 2 to 4 μm, while it reduces by 0.11 mV (27.5 %) when s is varied from 6 to 8 μm. Victim driver width (W_{n2}) is kept equal to 1 μm, and the corresponding PMOS width (W_{p2}) is 2.5 times W_{n2}.

The effect of interconnect length on peak negative and positive noise voltages is shown in Figs. 5.10 and 5.11, respectively. Line length is varied from 1 to 5 mm. It can be observed that peak noise increases with interconnect length. This is because the associated parasitic impedance parameters increase with the interconnect length. This increases the effective time constant of the victim driver, and consequently, peak noise increases. The effect of line-to-line spacing is also observed, and peak noise is seen to decrease with increased spacing. The proposed analytical model tracks the SPICE simulation results quite accurately. The maximum percentage error measured among all the observations is 5.75 and 6.32 %, respectively, for peak negative and positive noise voltages.

A comparison of the propagation delay of the aggressor driver and peak noise voltage based on the proposed analytical models and SPICE is presented in Table 5.2. It is seen that the maximum percentage error in the propagation delay using analytical model is 11.08 % with respect to SPICE, while the average percentage error is 4.03 %. The maximum and average percentage errors in peak coupling noise voltage determined by the proposed analytic model with respect to SPICE are 7.89 and 3.58 %, respectively. It is also observed that as the size of the quiet inverter is increased, the peak noise voltage is reduced. This can be illustrated by comparing the first and second rows of the Table 5.2. However, this technique increases the propagation delay of the active CMOS inverter. This is not a major concern since subthreshold circuits show low-speed performance.

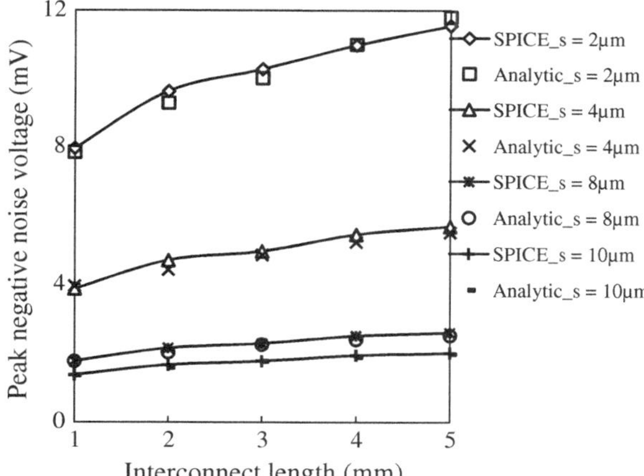

Fig. 5.10 Peak negative noise voltage with interconnect length

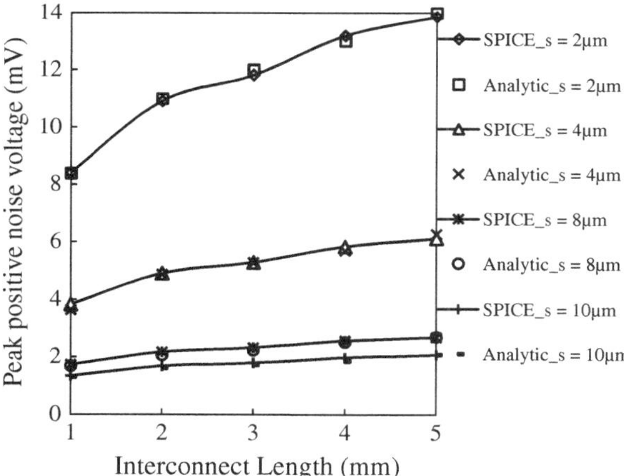

Fig. 5.11 Peak positive noise voltage with interconnect length

5.3.1 Power-Delay-Crosstalk-Product: Performance Criterion

Power-delay-crosstalk-product (PDCP) is a figure of merit to characterize the coupled interconnects performance when one inverter is active, while the other is quiet. The effect of subthreshold and super-threshold operation on PDCP is also investigated. The variations of PDC_+P with interconnect length and victim driver size in the aforementioned operating regimes are shown in Fig. 5.12. C_+ refers to the peak positive noise voltage.

It is observed that PDC_+P in subthreshold (st) is lower by an order compared to super-threshold. For example, in Fig. 5.12a, PDC_+P is 2.04 and 27.23 aJV in subthreshold and super-threshold regions, respectively, for 1 mm interconnect length. Increase in PDC_+P with interconnect length is expected since power

Table 5.2 Propagation delay and peak noise voltage for *Inv*1 active and *Inv*2 quiet

Aggressor parameters			Victim parameters			C_c (pf)	Initial state of *Inv*1	Delay of *Inv*1 (ns)		Error (%)	Peak voltage of *Inv*2 (mV)		Error (%)
W_{n1} (μm)	R_1 (Ω)	C_1 (pf)	W_{n2} (μm)	R_2 (Ω)	C_2 (pf)			SPICE	Analytic		SPICE	Analytic	
0.1	208.93	0.3	0.1	208.93	0.3	0.1	**Low-to-High**	128.47	114.7	10.72	−11.54	−11	4.70
0.1	208.93	0.3	1	208.93	0.3	0.1	**Low-to-High**	129.14	124.7	3.44	−1.24	−1.23	1.04
0.1	208.93	0.4	1	208.93	0.4	0.05	**Low-to-High**	129.81	121.6	6.32	−0.58	−0.53	7.89
0.1	208.93	0.4	0.1	208.93	0.4	0.05	**Low-to-High**	129.66	115.3	11.08	−5.68	−5.26	7.37
Aggressor parameters			Victim parameters			C_c (pf)	Initial state of *Inv*1	Delay of *Inv*1 (ns)		Error (%)	Peak Voltage of *Inv*2 (mV)		Error (%)
W_{p1} (μm)	R_1 (Ω)	C_1 (pf)	W_{n2} (μm)	R_2 (Ω)	C_2 (pf)			SPICE	Analytic		SPICE	Analytic	
0.2	208.93	0.3	0.1	208.93	0.3	0.1	**High-to-Low**	128.21	128.2	0.01	13.86	14.00	1.01
0.2	208.93	0.3	1	208.93	0.3	0.1	**High-to-Low**	129.05	128.7	0.27	1.32	1.30	1.06
0.2	208.93	0.4	1	208.93	0.4	0.05	**High-to-Low**	129.73	129.4	0.25	0.63	0.61	2.97
0.2	208.93	0.4	0.1	208.93	0.4	0.05	**High-to-Low**	129.56	129.4	0.12	6.11	6.27	2.56

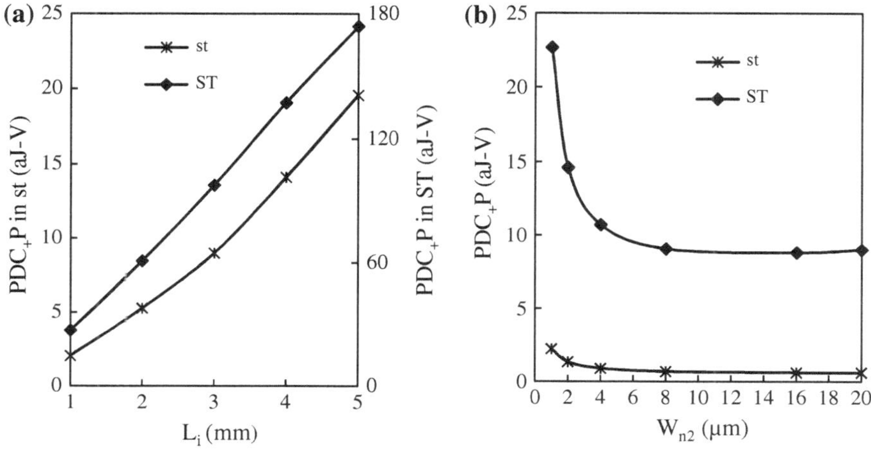

Fig. 5.12 Variation of PDC₊P with **a** interconnect length, **b** victim driver width

dissipation, propagation delay, and peak crosstalk increase with increase in interconnect parasitic impedance parameters.

It is seen from Fig. 5.12b that PDC₊P decreases with increasing victim driver width and then levels off beyond a certain width. For instance, PDC₊P in subthreshold shows a marginal variation beyond 2 µm victim driver width. It has been established earlier by the analysis that increasing victim driver width significantly lowers the peak crosstalk. This, however, does not much affect the propagation delay of the aggressor driver. On the other side, increasing victim driver width increases average power dissipation. The combined effect is that PDC₊P decreases by 59.9 % as victim driver width is increased up to 2 µm. Similar variations are

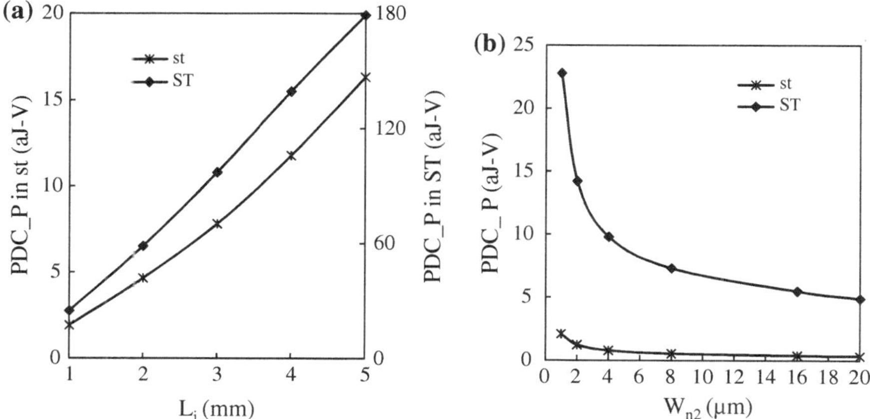

Fig. 5.13 Variation of PDC_P with **a** interconnect length, **b** victim driver width

observed for PDC_P and are shown in Fig. 5.13a, b. C_{-} refers to the peak negative noise voltage. For example, in Fig. 5.13a, PDC_P in subthreshold increases from 1.93 to 16.32 aJV as L_{i} is varied from 1 to 5 mm.

When victim driver width variations are considered as shown in Fig. 5.13b, PDC_P in subthreshold regime also shows nominal variations beyond 2 μm victim width. Thus, it is advantageous to keep interconnect and buffer dimensions limited to smaller values. Another conclusion of this analysis is that PDC_P in subthreshold (st) is lower by an order compared to super-threshold.

5.4 Concluding Remarks

In this chapter, the coupling noise behavior of capacitively coupled CMOS buffers in subthreshold regime is presented. Analytical expressions characterizing undershoot and overshoot at the output of quiet inverter have been developed. Expressions governing the output voltage and propagation delay of aggressor driver have also been derived. The effect of victim driver width and interconnect length on peak noise voltages is analyzed. The analytical models yield maximum percentage errors of 5.75 and 6.32 % for negative and positive noise voltages, respectively, among all the measured observations. The accuracy of these proposed models is also investigated under varying driver size and load conditions. Propagation delay estimates are within 5 %, while peak noise voltage is in 3.58 % average error with respect to SPICE. Thus, the proposed analytical models capture waveform shape, propagation delay, and noise peaks quite efficiently. Design techniques are also suggested to reduce the effect of coupling. It is shown that increasing line width, line-to-line spacing, and victim driver size are effective noise avoidance techniques. However, only a limited increase in the victim driver size and spacing is advantageous.

PDCP is defined as a figure of merit to characterize the performance of coupled interconnects. A lower value of PDCP indicates better performance. It is found out that that PDCP in subthreshold is lowered by an order compared to its super-threshold counterpart. This in turn shall facilitate ultra-low-power interconnect design.

Chapter 6
Variability in Subthreshold Interconnects

Keywords Monte Carlo · Process corners · Parametric analysis · Variability · Worst-case shift

As technology advances toward the nanometer regime, process variability has emerged as a serious concern in the design of VLSI circuits including interconnects. The process variations result in performance fluctuations in the circuit design and pose challenges as technology scales down. According to ITRS, scaled down VLSI technology together with novel process steps adds to the improvement and performance of deep submicron devices. However, fabrication process tolerances have not scaled proportionally with device dimensions. This has significantly increased the variation susceptibility in several key process parameters during the device fabrication. The increase in variability affects the design of low-power circuits in the nanometer regime. This causes fluctuations in the IC performance. Therefore, the relative impact of process variations on power and timing has become more significant with each technology generation.

The increased variation of process parameters of nanoscale devices not only results in higher average leakage but also causes a larger spread or standard variation of leakage power. Besides, temperature also affects the subthreshold system design and translates into exponential variations in the subthreshold current. Therefore, in deep submicron technologies, process, voltage, and temperature variations are becoming prominent factors affecting the design. Subsequently, this chapter focuses on the impact of PVT variations in the subthreshold interconnect performance.

6.1 Process Variability

Process variability can be classified as inter-die (die to die) and intra-die (within die). Inter-die or global variation refers to the variation from wafer to wafer or die to die on a same wafer. Traditionally, inter-die variations have become the main concern in CMOS digital circuit design. The inter-die variations constitute parameter deviations from the nominal value and originate from factors such as the processing temperature and equipment properties [229]. The intra-die variations are

R. Dhiman and R. Chandel, *Compact Models and Performance Investigations for Subthreshold Interconnects*, Energy Systems in Electrical Engineering,
DOI 10.1007/978-81-322-2132-6_6

random in nature and occur due to the semiconductor manufacturing process such as the random placement of dopant atoms in the channel region and channel length variations within a die [21]. However, intra-die variations have become just as important and their impact on frequency and power is becoming more and more pronounced. These variations are further classified into three categories which include device, interconnect and dynamic variations. It is therefore prudent to look at their variability impact on the performance constraints.

6.1.1 Device Variations

The process variation at the device level is related with physical geometric structure of the device. Device variations are fluctuations in MOS parameters such as gate length and oxide thickness during the device fabrication. Variations in the gate width are usually not considered since gate width is much larger than the gate length. The variations present at the transistor level are commonly known as front-end variations. These variations change the device threshold voltage and affect the circuit performance. Even with small V_T variations, drive currents of PMOS and NMOS can differ by an order of magnitude or even more in subthreshold circuits. As a consequence, the rise and fall times of output voltage differ significantly, thereby impacting the switching frequency and power dissipation. Variations in threshold voltage echoes for the propagation delay as well. The subthreshold circuits show a marked sensitivity toward threshold voltage variations and therefore jeopardize the circuit operation. The threshold voltage accounts for 30 % of the sources of variation in circuit performance. Threshold voltage variation has therefore always received a great deal of attention in the circuit design community. The impact of device parameters on V_T variations is discussed next.

(a) Effective Gate Length
Gate length variations in MOSFETs arise due to masking differences, etching process, spacer definition, and source/drain implantation. Of these, the primary sources of variation are the steps involved in the photolithographic and plasma etching processes which are considered as systematic variations and hence can be compensated. As gate length is reduced, the threshold voltage of short-channel device decreases. This is due to the closer proximity of source and drain areas whose surrounding depletion regions penetrate into a considerable portion of the channel. Therefore, less charge beneath the channel must be inverted by the gate voltage to reach the threshold voltage. The shift in n-channel threshold voltage (ΔV_T) originated by channel length scaling [230] is given as

$$\Delta V_T = [2(V_{bi} - \psi_s) + V_{ds}]\left(e^{-\frac{L_n}{2l}} + 2e^{-\frac{L_n}{l}}\right) \tag{6.1}$$

where V_{bi} is the built-in potential, ψ_s is the surface potential, and l is the characteristic length. This analytical approximation defines a short-channel effect known as the threshold voltage roll-off. As a result, any reduction in effective gate length increases the subthreshold leakage current and hence the power consumption.

(b) Oxide Thickness
The oxide thickness variation is due to the thin-film deposition process but is a relatively well-controlled parameter. Variations in the oxide thickness have a considerable effect on the threshold voltage since any variation in t_{ox} impacts the oxide capacitance. As a result, the variation in oxide thickness changes the subthreshold leakage current exponentially and can have a catastrophic impact on the power consumption in DSM technologies.

6.1.2 Interconnect Variations

Interconnects play significant role in determining the signal propagation, signal integrity, and power consumption of digital systems. Interconnect parameters include the width and thickness of metal line, spacing between interconnects, and inter-layer dielectric. Variations in interconnect geometry such as width and spacing arise due to the photolithographic and etching processes. Variations in the interconnect geometry result in change in the associated electrical parasitic impedance parameters. These electrical parameter variations therefore directly affect the performance of the circuit. The critical paths often contain long wires, and a good description of the interconnect geometry variation is needed for accurate circuit simulation. The variations in the several interconnection levels and dielectric layers are also known as back-end variations.

6.1.3 Dynamic Variations

Dynamic variations include supply voltage and temperature variations. The variability in supply voltage in the power grid occurs due to voltage drooping and currents being drawn by underlying devices. The variations of temperature which vary throughout the die are based on the location of high activity blocks.

6.2 Variability Analysis

It is necessary to understand and model manufacturing process variations for the prediction of device and circuit performance. This is because fluctuations in the semiconductor fabrication processes result in undesirable variations in the circuit performance. In order to analyze the effects of variability, three methods have been considered viz parametric, process corner, and Monte Carlo. The results obtained are presented in this section.

6.2.1 Parametric Analysis

The parametric analysis is suitable for studying the impact of individual process parameters on the circuit performance. In order to find out the parameter sensitivity, the parameters are varied one at the time in the range of $\pm 3\sigma$, keeping the rest of the values nominal. $\pm 3\sigma$ has been considered as the worst-case shift in the respective parameter value. The parameters varied are enlisted in Table 6.1 with their $\pm 3\sigma$ variations. These parameters have a normal distribution around their mean values.

In this analysis, 3σ variation of ± 12.5 % in the device threshold voltage, ± 15 % in the effective channel length, and ± 4 % in oxide thickness for both NMOS and PMOS devices are considered. ± 10 % variation in the supply voltage and ± 15 % in the interconnect resistance and capacitance per unit length are considered. The influence of these parameter variations on the circuit performance parameters for in-phase switching is derived for 65-nm technology. Comparison with SPICE simulations is also presented.

Expressions characterizing variability in the propagation delay and power dissipation as n-channel threshold voltage gets varied have been obtained using Eq. (4.32) and formulation carried out in [36], respectively, and are governed by (6.2) and (6.3), respectively, as

$$\Delta t_{\mathrm{P_{HL1}}}(V_{\mathrm{T}}) = \frac{\frac{\tau_{\mathrm{r}}}{V_{\mathrm{DD}}}\left(\mathrm{e}^{-V'_{\mathrm{T}}} - \mathrm{e}^{-V_{\mathrm{T}}}\right)}{\tau_{\mathrm{r}} + \frac{0.5V_{\mathrm{DD}}}{\gamma_{21}} - \frac{\tau_{\mathrm{r}}}{V_{\mathrm{DD}}}\left(1 - \mathrm{e}^{-V_{\mathrm{T}}}\right)} \tag{6.2}$$

$$\Delta P(V_{\mathrm{T}}) = \frac{B_{\mathrm{n_1}}\left[\mathrm{e}^{\left(\frac{V_{\mathrm{DD}} - V'_{\mathrm{T}}}{\eta_{\mathrm{n}} U_{\mathrm{th}}}\right)} - 1\right]}{f C_{\mathrm{eff}} V_{\mathrm{DD}} + \left(B_{\mathrm{n1}} + B_{\mathrm{p1}}\right)} \tag{6.3}$$

Table 6.1 Device parameters for 65-nm technology node as obtained by predictive technology model and their $\pm 3\sigma$ variations

Parameters	65-nm technology	
	NMOS (%)	PMOS (%)
V_{T} (V)	0.429 ± 12.5	−0.378 ± 12.5
L_{n} (L_{p}) (nm)	24.5 ± 15	24.5 ± 15
t_{ox} (nm)	1.2 ± 4	1.2 ± 4
V_{DD} (V)	0.36 ± 10	
R (Ω/mm)	21 ± 15	
C (fF/mm)	30 ± 15	

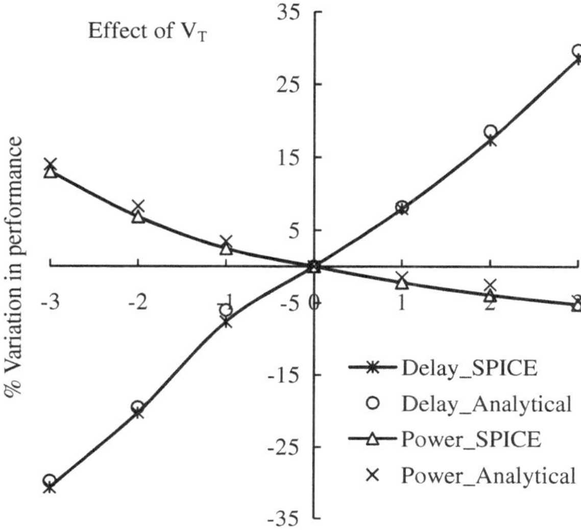

Fig. 6.1 Effect of threshold voltage variations on delay and power

Fluctuation in the n-channel transistor threshold voltage is represented by V'_T. C_{eff} is the output effective capacitance of aggressor driver switched per cycle and has been determined in accordance with [231]. The effect of threshold voltage variation on propagation delay and power dissipation can be seen in Fig. 6.1.

It is intuitive to learn from the plot that delay increases at a faster rate due to its exponential dependence on the threshold voltage. It is observed that an increase in the n-channel threshold voltage increases delay. Around ±29 % variation in delay is observed when threshold voltage is varied around its mean by $\pm 3\sigma$. Variance of power is less than that of delay with a maximum around 12 %. It may also be seen that delay and power have a negative correlation as threshold voltage varies and the variation in delay is higher than power for $\pm 3\sigma$ values.

Similar to threshold voltage variations, fluctuations in t_{ox} and L_n also affect the circuit performance. Expressions characterizing variability in the propagation delay and power dissipation as oxide thickness/effective channel length gets varied are governed by (6.4) and (6.5). Fluctuations in t_{ox} and L_n are contained in the terms B'_{n1} and B'_{n2}.

$$\Delta t_{P_{HL1}}(t_{\text{ox}}/L_n) = \frac{0.5V_{DD}\left[\frac{1}{\gamma'_{21}(C_2+C_c)B'_{n1}+C_cB'_{n2}} - \frac{1}{\gamma_{21}(C_2+C_c)B_{n1}+C_cB_{n2}}\right]}{\tau_r + \frac{0.5V_{DD}}{\gamma_{21}} - \frac{\tau_r}{V_{DD}}\left(1 - e^{-V_T}\right)} \tag{6.4}$$

$$\Delta P(t_{\text{ox}}/L_n) = \frac{\left[\left(B'_{p1} - B_{p1}\right) + \left(B'_{n1} - B_{n_1}\right)\right]}{fC_{\text{eff}}V_{DD} + \left(B_{n1} + B_{p1}\right)} \tag{6.5}$$

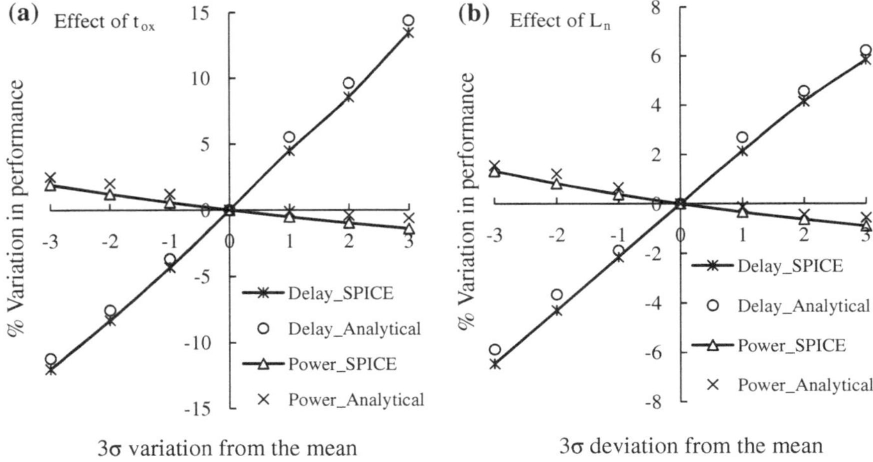

Fig. 6.2 The effect of t_{ox} (**a**) and L_n (**b**) variations on circuit delay and power

The effect of t_{ox} and L_n variation on circuit delay and power is seen in Fig. 6.2. Oxide thickness given in Table 6.1 is varied around its $\pm 3\sigma$ values, while the other parameters remained nominal. The variation in delay is higher than power dissipation for $\pm 3\sigma$ range of oxide thickness as can be seen in Fig. 6.2a. The variance in delay does not exceed ±14 %. Power dissipation varies to a lesser extent of ±2 %. Smaller variations in power dissipation may be attributed to the smaller variations in leakage power as oxide thickness gets varied around its mean value. The delay and power dissipation also exhibit negative correlation. The dependence of delay and power on the effective gate length variations is shown in Fig. 6.2b and follows the similar trend of t_{ox} variations. Delay shows a variation of around ±6 %, while maximum variance in power is 1.8 %.

The effect of supply voltage variations on propagation delay and power is discussed next. Compact analytical expressions governing variability in the propagation delay and power as supply voltage is varied have been derived using Eq. (4.32) and those provided in [35], respectively, as

$$\Delta t_{\mathrm{P_{HL1}}}(V_{\mathrm{DD}}) = \frac{\frac{0.5}{\gamma_{21}}\left(V'_{\mathrm{DD}} - V_{\mathrm{DD}}\right) + \tau_{\mathrm{r}}\left(\frac{1}{V_{\mathrm{DD}}} - \frac{1}{V'_{\mathrm{DD}}}\right)\left(1 - \mathrm{e}^{-V_{\mathrm{T}}}\right)}{\tau_{\mathrm{r}} + \frac{0.5V_{\mathrm{DD}}}{\gamma_{21}} - \frac{\tau_{\mathrm{r}}}{V_{\mathrm{DD}}}\left(1 - \mathrm{e}^{-V_{\mathrm{T}}}\right)} \tag{6.6}$$

$$\Delta P(V_{\mathrm{DD}}) = \frac{V'_{\mathrm{DD}}\left[B_{\mathrm{n1}}\mathrm{e}^{\left(\frac{V'_{\mathrm{DD}}-V_{\mathrm{T}}}{\eta_{\mathrm{n}}U_{\mathrm{th}}}\right)} + B_{\mathrm{p1}}\mathrm{e}^{\left(\frac{V'_{\mathrm{DD}}-V_{\mathrm{T}}}{\eta_{\mathrm{p}}U_{\mathrm{th}}}\right)}\right] - \left(B_{\mathrm{n1}} + B_{\mathrm{p1}}\right)V_{\mathrm{DD}}}{fC_{\mathrm{eff}}V_{\mathrm{DD}}^{2} + \left(B_{\mathrm{n1}} + B_{\mathrm{p1}}\right)V_{\mathrm{DD}}} \tag{6.7}$$

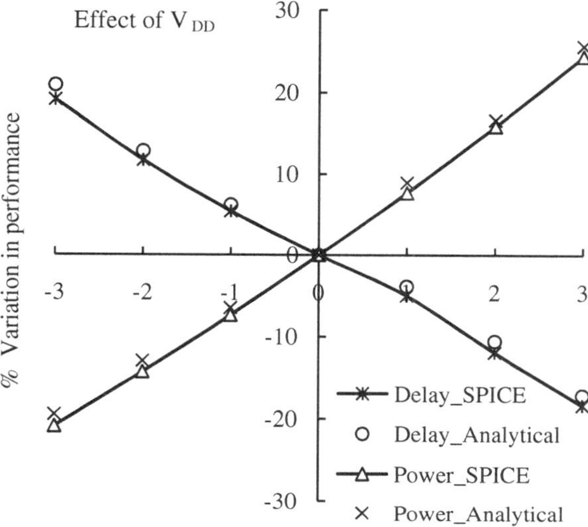

Fig. 6.3 Effect of supply voltage variations on delay and power

Fluctuations in the supply voltage are represented by V'_{DD}. Worst-case shift of 23.57 % in power and 20.21 % in delay are observed as supply voltage gets varied by ±10 % from its mean value. This can be seen in Fig. 6.3. It is also intuitive to learn from the plot that power dissipation increases as V_{DD} increases and increases at a faster rate due to the quadratic relation with V_{DD}.

When interconnect capacitance variation is considered, delay gets varied to a lesser extent of around ±5 % and power dissipation by ±5.9 % as shown in Fig. 6.4. This is expected as delay and power have a linear relationship with the interconnect capacitance. Interconnect resistance variation has not been considered, since in subthreshold, circuit performance is dictated by the driver resistance and device capacitance.

Analytical expressions governing variability in the propagation delay as load capacitance gets varied have been derived using Eq. (4.32) developed in Chap. 4 and given by,

$$\Delta t_{P_{HL1}} = \frac{0.5V_{DD}\left(\frac{\gamma'_{21}-\gamma_{21}}{\gamma_{21}\gamma'_{21}}\right)}{\tau_r + \frac{0.5V_{DD}}{\gamma_{21}} - \frac{\tau_r}{V_{DD}}\left(1 - e^{-V_T}\right)} \tag{6.8}$$

When variations in the interconnect coupling capacitance are considered, maxima and minima get varied to an extent of ±15 and ±12 % as shown in Fig. 6.5. This is because the peak noise voltage is directly proportional to the coupling capacitance.

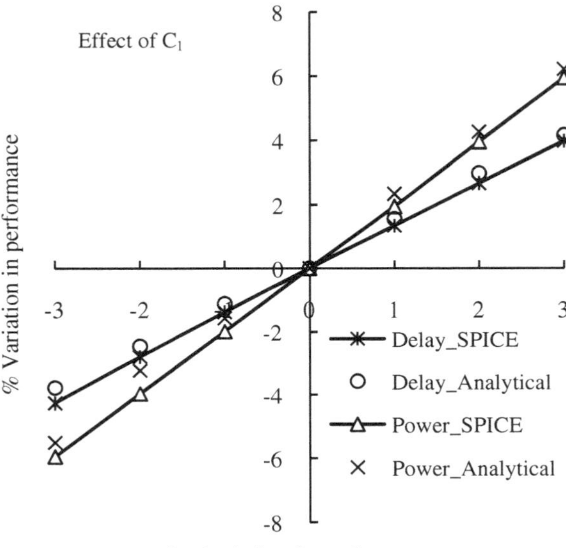

Fig. 6.4 Effect of interconnect capacitance variations on delay and power

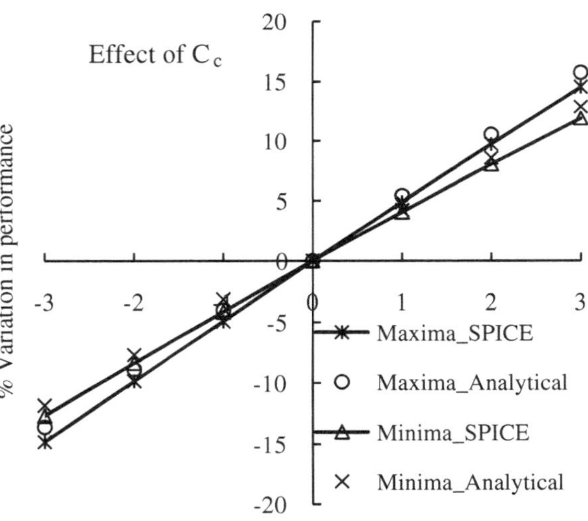

Fig. 6.5 Effect of interconnect coupling capacitance variations on maxima and minima

6.2.2 Process Corner Analysis

In this method, the process corners have been used to analyze the effect on the targeted requirements, i.e., power dissipation and delay. Using the combination of NMOS and PMOS, models for different process corners viz fast–fast (FF), slow–fast (SF), typical–typical (TT), fast–slow (FS), and slow–slow (SS) are considered where the first letter refers to the NMOS corner and the second letter refers to the PMOS corner. For example, a corner designated as FS denotes fast NMOS and slow PMOS. Fast and slow corners exhibit currents that are higher and lower than normal, respectively. These models are called process corners as they capture parameters that would make the circuit unusually fast or unusually slow.

The performance parameters of CMOS buffer-driven interconnect for different process corners at 65-nm technology nodes are shown in Fig. 6.6. Temperature variation is also analyzed. It may be seen from Fig. 6.6a that power dissipation increases with increase in temperature. For transistors having lowest threshold voltages, i.e., FF process corner, the variation of power dissipation with temperature is sharp as compared to other process corners. For example, at 100 °C, power dissipation in FF corner is 27.03 nW, while 9.91 nW in SS process corner. The increment in power dissipation with temperature is mainly due to the exponential sensitivity of subthreshold current with temperature.

Delay decreases with temperature for various process corners under consideration as shown in Fig. 6.6b. This is because of the reduction in threshold voltage with temperature and is 24.02 and 50.12 ns for FF and SS process corners at 25 °C. The FF process corner gives the minimum delay but consumes maximum power. Conversely, delay is maximum for SS corner and minimum for FF corner. The variations corresponding to other process corners lie in between these.

The performance analysis of two-line, three-line, and five-line coupled interconnects is discussed next. The two-line, three-line, and five-line coupled interconnects are identified as 2L, 3L, and 5L, respectively. The buffer-driven three and five coupled interconnects have been shown, respectively, in Figs. 6.7 and 6.8. C_1, C_2, C_3, C_4, and C_5 are the effective load capacitances of each CMOS inverter. C_c (C_{12}, C_{23}, C_{34}, and C_{45}) is the coupling capacitance between two neighboring lines. Out-of-phase switching is considered.

In Fig. 6.7, CMOS buffer driven by V_{in2} transitions dynamically opposite to the neighboring buffers. Likewise, in Fig. 6.8, CMOS buffer driven by V_{in3} transitions dynamically opposite to the neighboring buffers. These buffers or drivers in three- and five-line coupled interconnects are considered to be victim nets.

Figure 6.9 depicts the average power dissipation and delay at the victim net for various process corners at 130-nm technology node. It may be seen that SS process corner gives the highest power efficiency with slowest speed. For FF process corner, power dissipation is maximum as the currents are the highest. It is also observed that 2L shows the least power dissipation and 5L shows the highest power dissipation among the considered process corners. Similar trends are also observed for 90- and 65-nm technology nodes and are shown in Figs. 6.10 and 6.11 respectively. For example, in 90-nm technology, for FF corner, power dissipation is 61.66, 111.84, and 213.72 nW for 2L, 3L, and 5L, respectively. At 65-nm technology, the propagation delay is 31.61, 36.33, and 37.87 ns for 2L, 5L, and 3L, respectively. Thus, propagation delay is the highest for 3L and lowest for 2L. This is because effective victim wire capacitance to the ground is ($C + 2C_c$), ($C + 3C_c$), and ($C + 4C_c$) for 2L, 5L, and 3L, respectively, under out-of-phase stimulus [232].

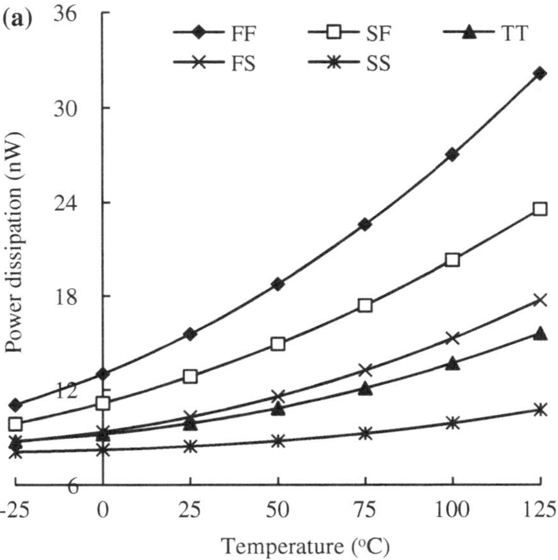

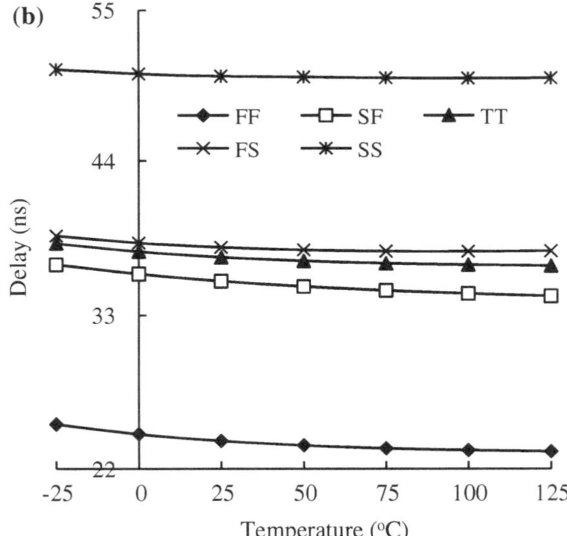

Fig. 6.6 Variation of **a** power dissipation and **b** delay with temperature for various process corners at 65-nm technology

6.2.3 Monte Carlo Analysis

The technique is based on iteratively evaluating the response of the deterministic model using sets of random numbers as inputs within certain specified ranges [233]. The typical–typical device model is used, and related parameters are varied using Gaussian distribution with $\pm 3\sigma$ deviation. To investigate the effects of process

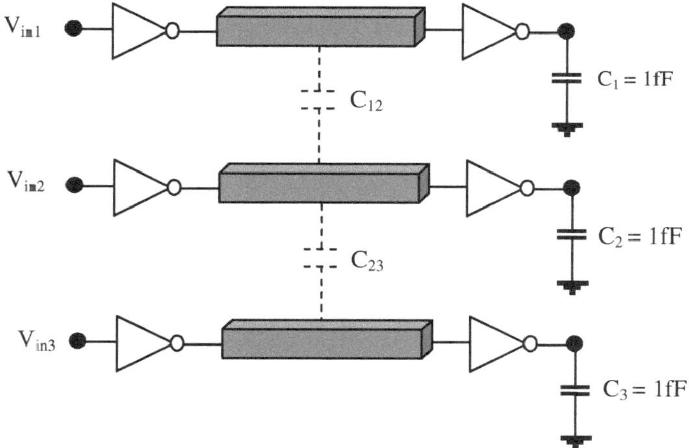

Fig. 6.7 Buffer-driven three coupled lines terminated by capacitive loads

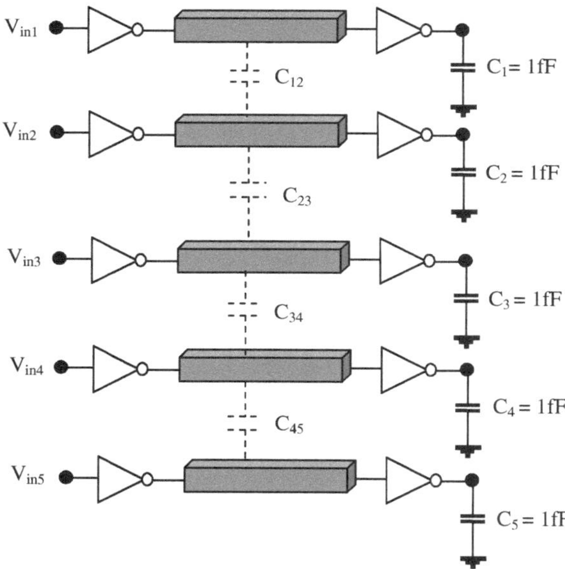

Fig. 6.8 Buffer-driven five coupled lines terminated by capacitive loads

variability on the subthreshold interconnect performance, Monte Carlo simulations have been performed for 1,000 runs at a temperature of 25 °C. Interconnect length of 5 mm has been taken. The Monte Carlo delay histogram for V_T variations with opposite phase switching is shown in Fig. 6.12.

It may be seen that delay for 5-mm line is normally distributed with peak at 37, 41, and 44 ns for 2L, 3L, and 5L coupled lines, respectively. Thus, greater delay variability is observed for 5L coupled interconnects. The delay and power

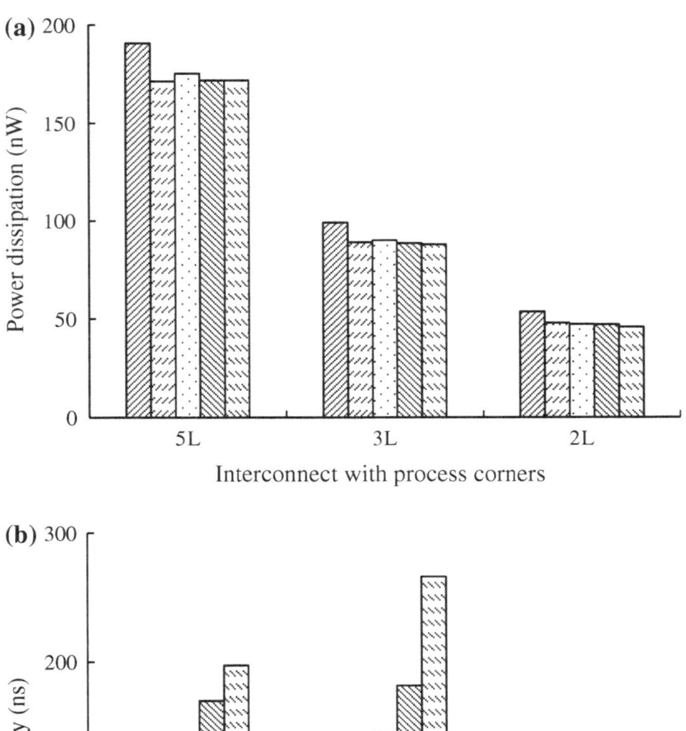

Fig. 6.9 Histograms of **a** power dissipation and **b** delay for 130-nm technology node

histogram for oxide thickness variation with opposite phase switching is shown in Fig. 6.13.

It is seen that for 68 runs, delay attains a value of 51 ns for two-line coupled structure. For three- and five-line coupled structures, delay is 51 and 56 ns for more than 60 and 50 runs, respectively. Propagation delay shows deviation from 31 to 70 ns, 33 to 77 ns, and 37 to 86 ns for 2L, 3L, and 5L coupled interconnect lines, respectively. Thus, five-line structure shows maximum delay variability. From

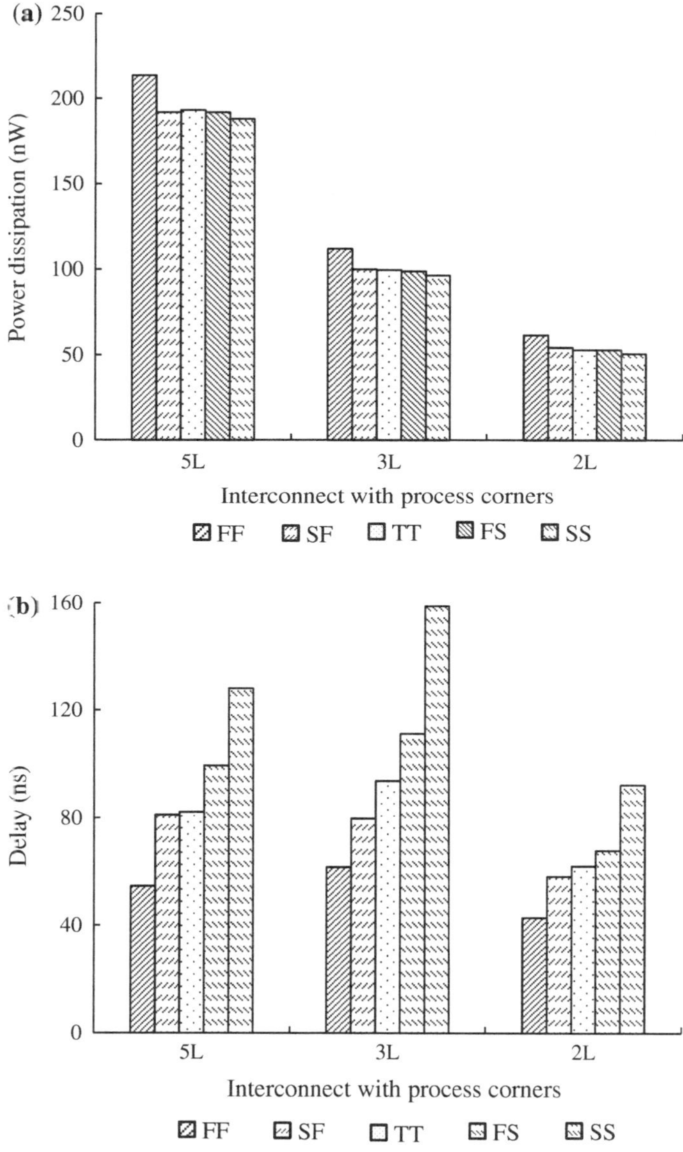

Fig. 6.10 Histograms of **a** power dissipation and **b** delay for 90-nm technology node

Fig. 6.13d, it is seen that power is also normally distributed with a peak of 91 nW for 189 runs. Power shows deviation from 87 to 104 nW.

Table 6.2 compares the delay distribution parameters among L_n, t_{ox}, and V_T. The greater sensitivity of delay against threshold voltage variations is 487.41 %. This is

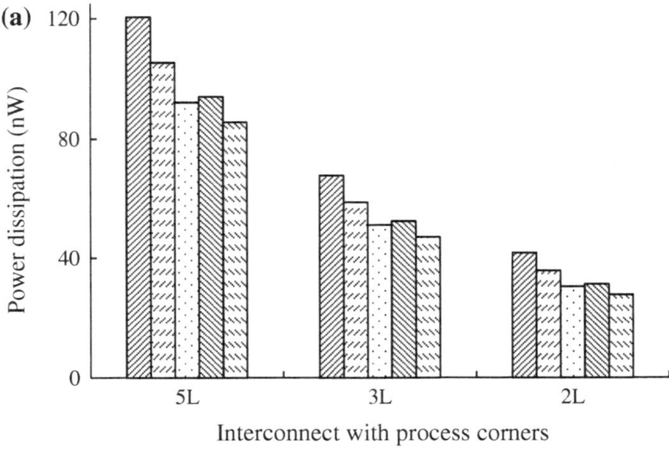

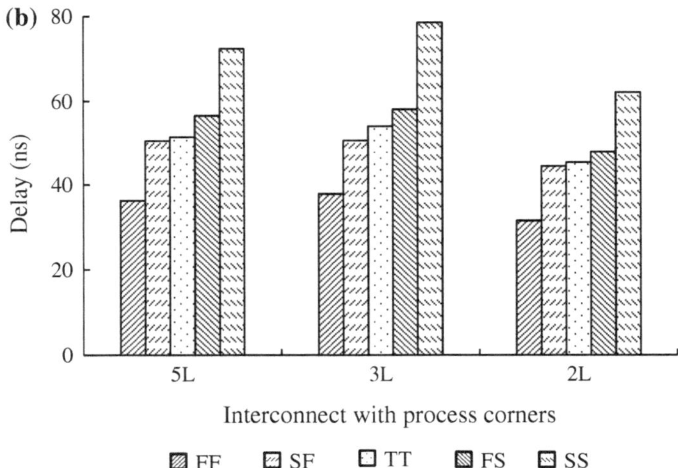

Fig. 6.11 Histograms of **a** power dissipation and **b** delay for 65-nm technology node

because of the exponential dependence of subthreshold bias current on the threshold voltage. L_n has the least effect, and quantitatively, it is 121.94 and 153.96 % for t_{ox} variations. Simultaneous variability in all the process variables shows a very large variation of 724.82 %. Despite the results being valuable and the accuracy being good in this technique, Monte Carlo technique is computationally expensive, especially when a large number of variables are involved as in studying the impact of parameter variation in semiconductor processes.

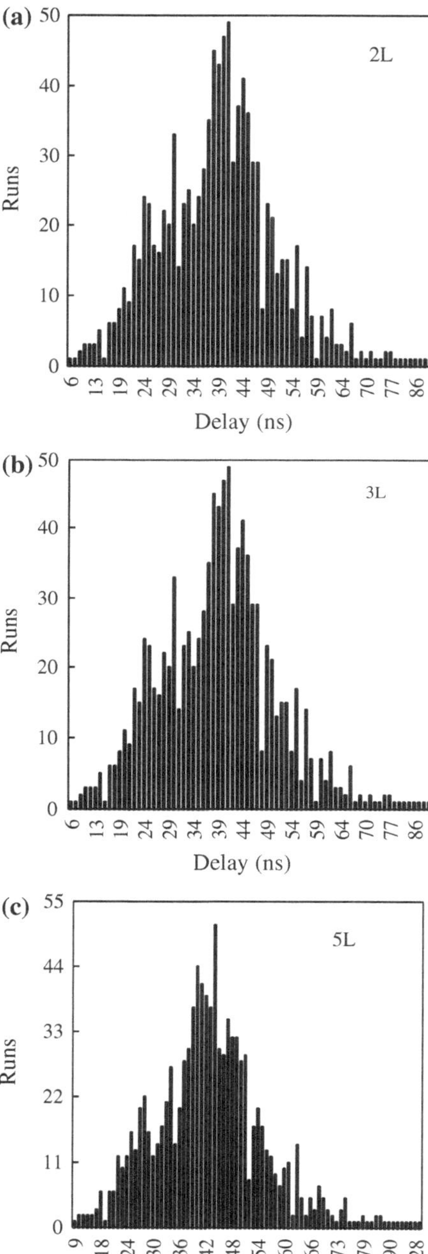

Fig. 6.12 Histogram of delay in center line with threshold voltage variations for **a** two-line interconnects, **b** three-line interconnects, and **c** five-line interconnects with 1,000 Monte Carlo runs

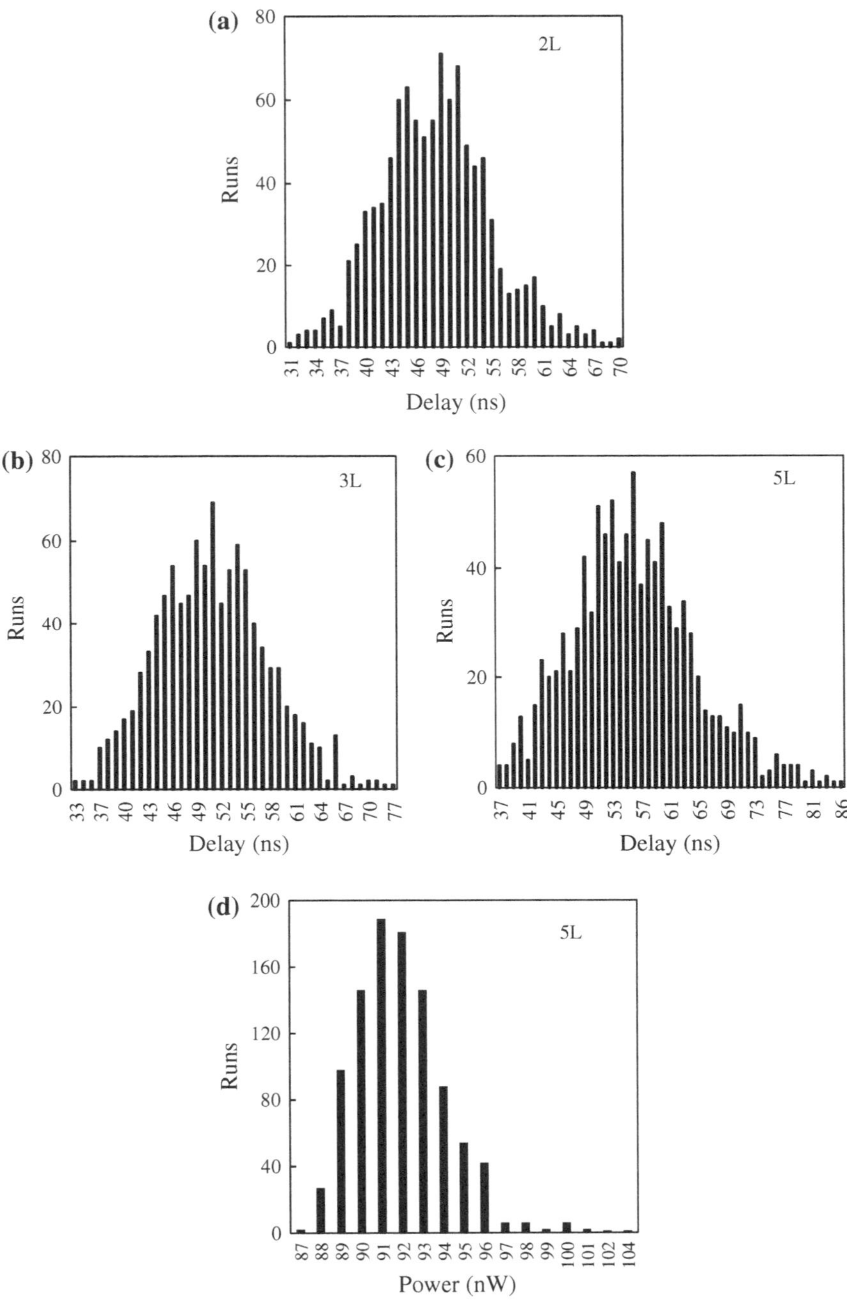

Fig. 6.13 Histogram of delay in center line with oxide thickness variations for **a** two-line interconnects, **b** three-line interconnects, **c** five-line interconnects, and **d** power histogram for five-line interconnects with 1,000 Monte Carlo Runs

Table 6.2 Delay in five-line coupled structure using Monte Carlo simulations for various process parameters

Process parameters	Nominal (ns)	Mean (ns)	Standard deviation (ns)	Maximum deviation from nominal	Maximum deviation per unit nominal (%)
L_n	55.60	55.58	3.80	67.8	121.94
t_{ox}	55.60	55.94	8.71	85.6	153.96
V_T	55.60	58.50	2.41	271	487.41
$L_{eff} + t_{ox} + V_T$	55.60	60.18	28.93	403	724.82

6.3 Effect of Temperature

It is well proven in the preceding chapters that interconnect delay, power dissipation, and crosstalk are the major showstoppers for performance improvement in ICs. However, along with these, the concern of thermal effects on VLSI interconnects is also emerging as another severe system design restriction [234, 235]. Temperature impacts the performance, reliability, and power consumption of integrated circuits.

The sources of temperature variation include ambient temperature and self-heating. Due to this, a circuit on the die can experience a wide temperature range of up to 0–70 °C for commercial applications and −55 to 125 °C for military applications. Temperature can vary across the die as well as with time as applications are run. Spatial temperature variations occur when some circuits are more active or denser, while others are less active. Temporal temperature variations occur when the amount of computation changes on the die are switching on or off, changing the power consumption. This in turn changes the temperature of the chip over time. The temperature is a strong function of the dissipated heat and its removal mechanisms. Keeping the overall operation temperature of a circuit low is consequently a desirable goal. This can be accomplished by limiting the power dissipation of the circuit and/or by using chip packages that support efficient heat removal [236]. Thus, it is essential to investigate the effects of temperature on delay and power dissipation of electronic circuits and interconnects.

6.3.1 Delay Variation with Temperature

The MOS transistor characteristics are strongly dependent on temperature. One of the main parameters responsible for this is the effective mobility, which decreases with temperature [237]. Effective mobility as a function of temperature (T) is given as

$$\mu(T) = \mu(T_r)(T/T_r)^{-\kappa_1} \tag{6.9}$$

where T_r is the room temperature in degrees Kelvin and κ_1 is a constant technology-dependent parameter. The value of κ_1 varies usually from 1.2 to 2. Other temperature-dependent parameters in MOS devices are the surface potential and the flat-band voltage.

These effects are manifested in the value of the threshold voltage as an almost straight-line decrease with temperature as shown in Eq. (6.10).

$$V_T(T) = V_T(T_r) - \kappa_2(T - T_r) \tag{6.10}$$

κ_2 is usually between 0.5 and 3 mV/K. Thus, a temperature increase tends to increase the drain current exponentially in the subthreshold region. Simultaneously, increase in temperature decreases the threshold voltage and hence delay [238]. However, it increases the power dissipation. The on-resistance of the device is also temperature dependent. Using the subthreshold current model, the on-resistance (R_{eqn}) of the device is given as

$$R_{eqn} = \frac{1}{\gamma_n} = \frac{1}{\mu_n C_{ox} \frac{W_n}{L_n} (\eta_n - 1) U_{th}} \tag{6.11}$$

The propagation delay as a function of temperature is obtained using Eq. (3.19) presented in Chap. 3 and is given by (6.12) as

$$t_{P_{HL}} = \tau_r + 0.5 V_{DD} \frac{C}{\mu_n C_{ox} \frac{W_n}{L_n} (\eta_n - 1) U_{th}^2} - \left(1 - e^{-\frac{V_{DD}}{\eta_n U_{th}}}\right)\left(RC + \frac{\tau_r \eta_n U_{th}}{V_{DD}}\right) \tag{6.12}$$

Delay variation with temperature using the proposed model and SPICE among various technology nodes is given in Table 6.3. It is observed that temperature rise results in delay decrement. A good agreement is seen between the SPICE and

Table 6.3 Analysis of propagation delay with temperature using SPICE and analytical approach

Temperature (°C)	Technology node								
	130 nm			90 nm			65 nm		
	Propagation delay (ns)								
	SPICE	Analytic	Error (%)	SPICE	Analytic	Error (%)	SPICE	Analytic	Error (%)
25	111.92	104.30	6.81	105.33	109.10	3.58	100.25	104.3	4.04
50	108.75	100.50	7.59	103.74	104.70	0.93	99.12	100.5	1.39
75	106.03	97.25	8.29	102.28	101.10	1.15	97.98	97.25	0.75
100	103.58	93.30	9.92	101.11	97.80	3.27	96.91	94.44	2.55
125	101.45	90.94	10.35	100.02	94.30	5.72	95.85	91.94	4.08
Average error (%)			8.59			2.93			2.56

analytical results. The average percentage errors are 8.59, 2.93, and 2.56 %, respectively, for 130-, 90-, and 65-nm technology nodes. Furthermore, as technology scales, delay improves. For example, SPICE extracted delay at 25 °C is 111.92, 105.33, and 100.25 ns in 130-, 90-, and 65-nm technology nodes.

6.4 Concluding Remarks

The principal device and interconnect parameters and their variation impact on circuit performance parameters viz power dissipation, delay, and crosstalk have been investigated. The essential parameters to account for variations are threshold voltage, oxide thickness, and effective channel length. The different approaches and methodologies used to handle the impact of process variations in circuit design are also examined. The variability analysis has been carried out using parametric, process corner, and Monte Carlo techniques. Variations of the parameters are given with a sigma sweep using parametric and Monte Carlo analysis.

Analytical expressions characterizing variability based on the parametric analysis have been developed. Variability estimates based on the analytical expressions are within 10 % compared to SPICE. It is shown that delay shows a marked sensitivity toward threshold voltage variations. For threshold voltage variations, the variance of delay is ±29 %, while the worst-case shift in power dissipation is around 12 %. Power is relatively constant for other device parameter $\pm 3\sigma$ variations with maximum variance not exceeding ±2 %. However, power is most sensitive to variations in the supply voltage and produce worst-case shift of 23.57 % as supply voltage gets varied by ±10 % from its mean value. Variations in interconnect coupling capacitance are also considered and produce ±15 and ±12 % variance in maxima and minima, respectively. Different process corners have been used to analyze the effect on the performance of two-line, three-line, and five-line CMOS buffer-driven coupled interconnects. Five-line and three-line structures show maximum variability in terms of power dissipation and delay. Monte Carlo analysis shows that there is a significant increase in delay when $\pm 3\sigma$ variations in the threshold voltage occur.

The impact of temperature variations is investigated to analyze the thermal effects in long interconnects. Temperature rise results in the delay decrement. It is also observed that with every 25 °C rise in temperature, leakage power nearly doubles for deep submicron technologies.

Bibliography

1. J.M. Rabaey, A. Chandrakasan, B. Nikolic, *Digital Integrated Circuits: A Design Perspective* (Pearson Education, Upper Saddle River, 2003)
2. N. Sherwani, *Algorithms for VLSI Physical Design and Automation* (Springer, New Delhi, 2005)
3. C.J. Uchibori, M. Lee, X. Zhang, P.S. Ho, T. Nakamura, Impact of Cu/low-k interconnect design on chip package interaction in flip chip package, in *Proceedings of the American Institute of Physics* (2008), pp. 185–196
4. International Technology Roadmap for Semiconductors (ITRS) (2009), http://public.itrs.net
5. J.A. Davis, R. Venkatesan, A. Kaloyeros, M. Beylansky, S.J. Souri, K. Banerjee, K.C. Saraswat, A. Rahman, R. Reif, J.D. Meindl, Interconnect limits on giga scale integration (GSI) in the 21st century, in *Proceedings of the IEEE* (2001), pp. 305–332
6. R. Chandel, S. Sarkar, R.P. Agarwal, Repeater insertion in global interconnects in VLSI circuits. Microelectron. Int. **22**, 43–50 (2005)
7. D. Duarte, V. Narayanan, M.J. Irwin, Impact of technology scaling in the clock system power, in *Proceedings of the IEEE Computer Society Annual Symposium VLSI* (2002), pp. 52–57
8. E. Barke, Line-to-ground capacitance calculation: a comparison. IEEE Trans. Comput. Aided Des. Integr. Circuits Syst. **17**, 295–298 (1998)
9. L. Gal, On-chip crosstalk-The new signal integrity challenge, in *Proceedings of the IEEE Custom Integrated Circuits Conference* (1995), pp. 251–254
10. B.K. Kaushik, S. Sarkar, R.P. Agarwal, R.C. Joshi, Crosstalk analysis and repeater insertion in crosstalk aware coupled VLSI interconnects. Microelectron. Int. **23**, 55–63 (2008)
11. A. Wang, B.H. Calhoun, A.P. Chandrakasan, *Sub-threshold Design for Ultra Low-Power Systems* (Springer, New York, 2006)
12. H. Soeleman, K. Roy, Ultra-low power digital subthreshold logic circuits, in *Proceedings of the International Symposium Low Power Electronics and Design* (1999), pp. 94–96
13. H. Soeleman, K. Roy, Digital CMOS logic operation in the sub-threshold region, in *Proceedings of the International Symposium VLSI Design* (2000), pp. 107–112
14. C.H.I. Kim, H. Soeleman, K. Roy, Ultra-low-power DLMS adaptive filter for hearing aid applications. IEEE Trans. Very Large Scale Integr. Syst. **11**, 1058–1067 (2003)
15. B. Zhai et al., Energy-efficient subthreshold processor design. IEEE Trans. Very Large Scale Integr. Syst. **17**, 1127–1137 (2009)
16. A. Wang, A. Chandrakasan, A 180mV FFT processor using subthreshold circuit techniques, in *International Solid-State Circuits Conference Digest of Technical Papers* (2004), pp. 292–529
17. T. Kim, J. Liu, J. Keane, C. Kim, A high-density subthreshold SRAM with data-independent bitline leakage and virtual ground replica scheme, in *IEEE International Solid-State Circuits Conference Digest of Technical Papers* (2007), pp. 330–331

R. Dhiman and R. Chandel, *Compact Models and Performance Investigations for Subthreshold Interconnects*, Energy Systems in Electrical Engineering,
DOI 10.1007/978-81-322-2132-6

18. M. Seok, S. Hanson, Y.S. Lin, Z. Foo, D. Kim, Y. Lee, N. Liu, D. Sylvester, D. Blaauw, The phoenix processor: a 30pW platform for sensor applications, in *IEEE Symposium VLSI Circuits* (2008), pp. 188–189
19. D. Markovic, C.C. Wang, L.P. Alarcon, T.-T Liu, J.M. Rabaey, Ultra low-power design in near-threshold region, in *Proceedings of the IEEE* (2010), pp. 237–252
20. B.H. Calhoun, J.F. Ryan, S. Khanna, M. Putic, Flexible circuits and architecture for ultra low power, in *Proceedings of the IEEE* (2010), pp. 267–281
21. O.S. Unsal, J.W. Tschanz, K. Bowman, V. De, X. Vera, A. Gonzlez, O. Ergin, Impact of parameter variations on circuits and microarchitecture, *IEEE Micro*, vol. 26 (2006), pp. 30–39
22. S.K. Gupta, A. Raychowdhary, K. Roy, Digital computation in subthreshold region for ultra low-power operation: a device-circuit-architecture codesign perspective, in *Proceedings of the IEEE* (2010), pp. 160–190
23. J. Kil, J. Gu, C.H. Kim, A high-speed variation-tolerant interconnect technique for sub-threshold circuits using capacitive boosting. IEEE Trans. Very Large Scale Integr. Syst. **16**, 456–465 (2008)
24. Y. Ho, H.K. Chen, C. Su, Energy-effective sub-threshold interconnect design using high-boosting predrivers. IEEE J. Emerg. Sel. Top. Circuits Syst. **2**, 307–312 (2012)
25. Y. Ismail, E.G. Friedman, Effects of inductance on the propagation delay and repeater insertion in VLSI circuits. IEEE Trans. Very Large Scale Integr. Syst. **8**, 195–206 (2000)
26. C. Guoqing, E.G. Friedman, Low power repeaters driving RC and RLC interconnects with delay and bandwidth constraints. IEEE Trans. Very Large Scale Integr. Syst. **14**, 161–172 (2006)
27. D. Sylvester, C. Hu, S. Nakagawa, S.Y. Oh, Interconnect scaling: signal integrity and performance in future high-speed CMOS designs, in *IEEE Symposium VLSI Technology, Digest of Technical Papers* (1998), pp. 42–43
28. S. Asai, Y. Wada, Technology challenges for integration near and below 0.1μm, in *Proceedings of the IEEE* (1997), pp. 505–520
29. P.K. Bondyopadhyay, Moore's law governs the silicon revolution, in *Proceedings of the IEEE* (1998), pp. 78–81
30. H.P. Wong, D.J. Frank, P.M. Solomon, C.J. Wann, J. Welser, Nanoscale CMOS, in *Proceedings of the IEEE* (1999), pp. 537–570
31. R. Ho, K.W. Mai, M.A. Horowitz, The future of wires, in *Proceedings of the IEEE* (2001), pp. 490–504
32. R.W. Keyes, Fundamental limits of silicon technology, in *Proceedings of the IEEE* (2001), pp. 227–238
33. J.T. Kong, CAD for nanometer silicon design challenges and success. IEEE Trans. Very Large Scale Integr. Syst. **12**, 1132–1147 (2004)
34. D. Sylvester, C. Hu, Analytical modeling and characterization of deep submicrometer interconnect, in *Proceedings of the IEEE* (2001), pp. 634–664
35. A. Wang, A.P. Chandrakasan, S.V. Kosonocky, Optimal supply and threshold scaling for subthreshold CMOS circuits, in *Proceedings of the IEEE Computer Society Annual Symposium on VLSI* (2002), pp. 5–9
36. B.C. Paul, A. Raychowdhury, K. Roy, Device optimization for digital subthreshold logic operation. IEEE Trans. Electron Devices **52**, 237–247 (2005)
37. R.H. Havemann, J.A. Hutchby, High-performance interconnects: an integration overview, in *Proceedings of the IEEE* (2001), pp. 586–601
38. R. Achar, M.S. Nakhla, Simulation of high-speed interconnects, in *Proceedings of the IEEE* (2001), pp. 693–728
39. A. Naeemi, J.A. Davis, J.D. Meindl, Compact physical models for multilevel interconnect crosstalk in gigascale integration (GSI). IEEE Trans. Electron Devices **51**, 1902–1912 (2004)
40. Y. Taur et al., CMOS scaling into the nanometer regime, in *Proceedings of the IEEE* (1997), pp. 486–504

41. R. Chandel, Study of voltage-scaled repeaters for long interconnects in VLSI circuits, Ph.D. Dissertation, IIT Roorkee, India, 2005
42. Y. Eo, W.R. Eisenstadt, High speed VLSI interconnect modeling based on S-parameter measurements. IEEE Trans. Compon. Hybrids Manufact. Technol. **16**, 555–562 (1993)
43. J. Qian, S. Pullela, L. Pillage, Modeling the effective capacitance for the RC interconnect of CMOS gates. IEEE Trans. Comput. Aided Des. Integr. Circuits Syst. **13**, 1526–1535 (1994)
44. N. Delorme, M. Belleville, J. Chilo, Inductance and capacitance analytic formulas for VLSI interconnects. Electron. Lett. **32**, 996–997 (1996)
45. F. Moll, M. Roca, A. Rubio, Inductance in VLSI interconnection modeling, in *IEEE Proceedings Circuits, Devices and Systems* (1998), pp. 175–179
46. S.C. Wong, T.G.Y. Lee, D.J. Ma, C.J. Chao, An empirical three dimensional crossover capacitance model for multilevel interconnect VLSI circuits. IEEE Trans. Semicond. Manuf. **13**, 219–227 (2000)
47. N.D. Arora, K.V. Raol, R. Schumann, L.M. Richardson, Modeling and extraction of interconnect capacitances for multilayer VLSI circuits. IEEE Trans. Comput. Aided Des. Integr. Circuits Syst. **15**, 58–67 (1996)
48. W. Jin, Y. Eo, W.R. Eisenstadt, J. Shim, Fast and accurate quasi three-dimensional capacitance determination of multilayer VLSI interconnects. IEEE Trans. Very Large Scale Integr. Syst. **9**, 450–460 (2001)
49. S.P. Sim, S. Krishnan, D.M. Petranovic, N.D. Arora, K. Lee, C.Y. Yang, A unified RLC model for high speed on-chip interconnects. IEEE Trans. Electron Devices **50**, 1501–1510 (2003)
50. X. Huang, P. Restle, T. Bucelot, Y. Cao, T.J. King, C. Hu, Loop-based interconnect modeling and optimization approach for multi gigahertz clock network design. IEEE J. Solid-State Circuits **38**, 457–463 (2003)
51. E.B. Rosa, The self and mutual inductances of linear conductors. Bull. Bur. Stan. **4**, 301–344 (1998)
52. K. Banerjee, A. Mehrotra, Analysis of on-chip inductance effects for distributed RLC interconnects. IEEE Trans. Comput. Aided Des. Integr. Circuits Syst. **21**, 904–915 (2002)
53. H. Ymeri, B. Nauwelaers, K. Maex, Frequency-dependent mutual resistance and inductance formulas for coupled IC interconnects on an Si-SiO_2 substrate. Integr. VLSI J. **30**, 133–141 (2001)
54. J. Cong, Z. Pan, Interconnect performance estimation models for design planning. IEEE Trans. Comput. Aided Des. Integr. Circuits Syst. **20**, 739–752 (2001)
55. S.K. Ip, Time-domain vector fitting extracted characteristic model for VLSI interconnects analysis, in *Proceedings of the Asia-Pacific Microwave Conference* (2005), pp. 958–962
56. W.C. Elmore, The transient response of damped linear networks with particular regard to wide-band amplifiers. J. Appl. Phys. **19**, 55–63 (1948)
57. S.S. Sapatnekar, RC interconnect optimization under the Elmore delay model, in *31st Design Automation Conference* (1994), pp. 387–391
58. R. Gupta, B. Tutuianu, L.T. Pillage, The Elmore delay as a bound for RC trees with generalized input signals. IEEE Trans. Comput. Aided Des. Integr. Circuits Syst. **16**, 95–104 (1997)
59. L.M. Brocco, S.P. McCormick, J. Allen, Macromodeling CMOS circuits for timing simulation. IEEE Trans. Comput. Aided Des. Integr. Circuits Syst. **7**, 1237–1249 (1988)
60. R.O. Brien, T.L. Savarino, Modeling the driving-point characteristic of resistive interconnect for accurate delay estimation, in *IEEE International Conference on Computer-Aided Design* (1989), pp. 512–515
61. T. Sakurai, Closed-form expressions for interconnection delay, coupling and crosstalk in VLSIs. IEEE Trans. Electron Devices **40**, 118–124 (1993)
62. A.B. Kahng, S. Muddu, An analytical delay model for RLC interconnects. IEEE Trans. Comput. Aided Des. Integr. Circuits Syst. **16**, 1507–1514 (1997)

63. J.A. Davis, J.D. Meindl, Is interconnect the weak link? IEEE Circuits Devices Mag. **14**, 30–36 (1998)
64. J.A. Davis, V.K. De, J.D. Meindl, A stochastic wire length distribution for gigascale integration (GSI)-Part II: applications to clock frequency, power dissipation, and chip size estimation. IEEE Trans. Electron Devices **45**, 590–597 (1998)
65. H.G. Brachtendorf, R. Laur, An accurate model for the transient simulation of lossy interconnects based on a novel discretization formula. Integr. VLSI J. **29**, 117–129 (2000)
66. E. Chiprout, Interconnect and substrate modeling and analysis: An overview. IEEE J. Solid-State Circuits **33**, 1445–1452 (1998)
67. D. Pamunuwa, H.Tenhumen, Repeater insertion to minimize delay in coupled interconnects, in *International Conference on VLSI Design* (2001), pp. 513–517
68. J.A. Davis, J.D. Meindl, Compact distributed RLC interconnect models-Part I: single line transient, time delay, and overshoot expressions. IEEE Trans. Electron Devices **47**, 2068–2077 (2000)
69. J.A. Davis, J.D. Meindl, Compact distributed RLC interconnect models-Part II: coupled line transient expressions and peak crosstalk in multilevel interconnect networks. IEEE Trans. Electron Devices **47**, 2078–2087 (2000)
70. R. Venkatesan, J.A. Davis, J.D. Meindl, Compact distributed RLC interconnect models-Part III: transients in single and coupled lines with capacitive load termination. IEEE Trans. Electron Devices **50**, 1081–1093 (2003)
71. R. Venkatesan, J.A. Davis, J.D. Meindl, Compact distributed RLC interconnect models-Part IV: unified models for time delay, crosstalk, and repeater insertion. IEEE Trans. Electron Devices **50**, 1094–1102 (2003)
72. Q. Xu, P. Mazumder, Equivalent-circuit interconnect modeling based on the fifth order differential quadrature methods. IEEE Trans. Very Large Scale Integr. Syst. **11**, 1068–1079 (2003)
73. A. Maheshwari, W. Burleson, Differential current-sensing for on-chip interconnects. IEEE Trans. Very Large Scale Integr. Syst. **12**, 1321–1329 (2004)
74. B. Chen, H. Yang, R. Luo, H. Wang, A novel method for worst case interconnect delay estimation. IEEE Trans. Circuits Syst. I Fundam. Theory Appl. **50**, 778–781 (2003)
75. R. Singhal, G. Choi, R.N. Mahapatra, Data handling limits of on-chip interconnects. IEEE Trans. Very Large Scale Integr. Syst. **16**, 707–713 (2008)
76. T. Lehtonen, D. Wolpert, P. Liljeberg, J. Plosila, P. Ampadu, Self-adaptive system for addressing permanent errors in on-chip interconnects. IEEE Trans. Very Large Scale Integr. Syst. **18**, 527–540 (2010)
77. A. Morgenshtein, E.G. Friedman, R. Ginosar, A. Kolodny, Unified logical effort-A method for delay evaluation and minimization in logic paths with RC interconnect. IEEE Trans. Very Large Scale Integr. Syst. **18**, 689–696 (2010)
78. W. Shockley, A unipolar field effect transistor, in *Proceedings of the IRE* (1952), pp. 1365–1376
79. H. Shichman, D.A. Hodges, Modeling and simulation of insulated gate field effect transistor switching circuits. IEEE J. Solid-State Circuits **SC-3**, 285–289 (1968)
80. T. Sakurai, A.R. Newton, Alpha power law MOSFET model and its applications to CMOS inverter delay and other formulas. IEEE J. Solid-State Circuits **25**, 584–594 (1990)
81. A.C. Deng, Y.C. Shiau, Generic linear RC delay modeling for digital CMOS circuits. IEEE Trans. Comput. Aided Des. Integr. Circuits Syst. **9**, 367–376 (1990)
82. M.C. Chung, J.E. Moon, K.O. Ping-Keung, C. Hu, Performance and reliability design issues for deep-submicrometer MOSFET's. IEEE Trans. Electron Devices **38**, 545–554 (1991)
83. S. Dutta, S.S.M. Shetti, S.L. Lusky, A comprehensive delay model for CMOS inverters. IEEE J. Solid-State Circuits **30**, 864–871 (1995)
84. L. Bisdounis, S. Nikolaidis, O. Koufopavlou, Analytical transient response and propagation delay evaluation of the CMOS inverter for short channel devices. IEEE J. Solid-State Circuits **33**, 302–306 (1998)

85. A. Nabavi-Lishi, N.C. Rumin, Inverter models of CMOS gates for supply current and delay evaluation. IEEE Trans. Comput. Aided Des. Integr. Circuits Syst. **13**, 1271–1279 (1994)
86. A. Hirata, H. Onodera, K. Tamaru, Estimation of propagation delay considering short-circuit current for static CMOS gates. IEEE Trans. Circuits Syst. I Fundam. Theory Appl. **45**, 1194–1198 (1998)
87. M. Pattanaik, S. Banerjee, B.K. Bahinipati, Power delay optimization of nanoscale CMOS inverter using geometric programming. WSEAS Trans. Circuits Syst. **5**, 536–541 (2006)
88. J.M. Daga, D. Auvergne, A comprehensive delay macro modeling for sub micrometer CMOS logics. IEEE J. Solid-State Circuits **34**, 42–55 (1999)
89. T. Raja, V.D. Agrawal, M.L. Bushnell, Variable input delay CMOS logic for low power design. IEEE Trans. Very Large Scale Integr. Syst. **17**, 1534–1545 (2009)
90. S.H.K. Embabi, R. Damodaran, Delay models for CMOS, BiCMOS, BiNMOS circuits and their applications for timing simulations. IEEE Trans. Comput. Aided Des. **13**, 1132–1142 (1994)
91. F. Moll, M. Roca, *Interconnection Noise in VLSI Circuits* (Kluwer, New York, 2004)
92. K.T. Tang, E.G. Friedman, Delay and noise estimation of CMOS logic gates driving coupled resistive-capacitive interconnections. Integr. VLSI J. **29**, 131–165 (2000)
93. W. Chen, S.K. Gupta, M.A. Breuer, Analytic models for crosstalk delay and pulse analysis under non-ideal inputs, in *Proceedings of the International Test Conference* (1997), pp. 809–818
94. A.B. Kahng, S. Muddu, D. Vidhani, Noise and delay uncertainty studies for coupled RC interconnects, in *Proceedings of the 12th IEEE International ASIC/SOC Conference* (1999), pp. 3–8
95. D.H. Xie, M. Nakhla, Delay and crosstalk simulation of high-speed VLSI interconnects with nonlinear terminations. IEEE Trans. Comput. Aided Des. Integr. Circuits Syst. **12**, 1798–1811 (1993)
96. J. Poltz, Determining noise levels in VLSI circuits, in *IEEE International Symposium Electromagnetic Compatibility* (1993), pp. 340–345
97. M. Kuhlmann, S. Sapatnekar, K. Parhi, Efficient crosstalk estimation, in *International Conference Computer Design* (1999), pp. 266–272
98. A. Vittal, L.H. Chen, M.M. Sadowska, K.P. Wang, S. Yang, Crosstalk in VLSI interconnections. IEEE Trans. Comput. Aided Des. Integr. Circuits Syst. **18**, 1817–1824 (1999)
99. K.T. Tang, E.G. Friedman, Interconnect coupling noise in CMOS VLSI circuits, in *Proceedings of the ACM International Symposium Physical Design* (1999), pp. 48–53
100. H. Kawaguchi, T. Sakurai, Delay and noise formulas for capacitively coupled distributed RC lines, in *Proceedings of the Design Automation Conference* (1998), pp. 35–43
101. L. Ling, D. Blaauw, P. Mazumder, Accurate crosstalk noise modeling for early signal integrity analysis. IEEE Trans. Comput. Aided Des. Integr. Circuits Syst. **22**, 627–634 (2003)
102. A. Devgan, Efficient coupled noise estimation for on-chip interconnects, in *IEEE/ACM International Conference Computer-Aided Design, Digest of Technical Papers* (1999), pp. 1817–1824
103. P. Heydari, M. Pedram, Capacitive coupling noise in high-speed VLSI circuits. IEEE Trans. Comput. Aided Des. Integr. Circuits Syst. **24**, 478–488 (2005)
104. M. Shoji, *Theory of CMOS Digital Circuits and Circuit Failures* (Princeton University Press, Princeton, 1992)
105. M. Hashimoto, Y. Yamada, H. Onodera, Capturing crosstalk induced waveform for accurate static timing analysis, in *Proceedings of the International Symposium Physical Design* (2003), pp. 18–23
106. Y. Eo, W.R Eisenstadt, J.Y. Jeong, O.K. Kwon, A new on-chip interconnect crosstalk model and experimental verification for CMOS VLSI circuit design. IEEE Trans. Electron Devices **47**, 129–140 (2000)

107. M.R. Becer, D. Blaauw, V. Zolotov, R. Panda, I.N. Hajj, Analysis of noise avoidance techniques in DSM interconnects using a complete crosstalk noise model, in *Proceedings of the Design, Automation and Test Conference* (2002), pp. 456–463
108. S. Hasan, A.K. Palit, W. Anheier, Equivalent victim model of the coupled interconnects for simulating crosstalk induced glitches and delays, in *IEEE Workshop Signal Propagation on Interconnects* (2009), pp. 1–4
109. S. Tuuna, L.R. Zheng, J. Isoaho, H. Tenhunen, Modeling of on-chip bus switching current and its impact on noise in power supply grid. IEEE Trans. Very Large Scale Integr. Syst. **16**, 766–770 (2008)
110. P. Bazargan-Sabet, P. Renault, An event-driven approach to crosstalk noise analysis (digital ICs), in *IEEE 36th Annual Simulation Symposium* (2003), pp. 319–326
111. B.K. Kaushik, S. Sarkar, R.P. Agarwal, Width optimization of global inductive VLSI interconnects. Microelectron. Int. **23**, 26–30 (2006)
112. K. Agarwal, D. Sylvester, D. Blaauw, Modeling and analysis of crosstalk noise in coupled RLC interconnects. IEEE Trans. Comput. Aided Des. Integr. Circuits Syst. **25**, 892–901 (2006)
113. L.H. Chen, M.M. Sadowska, Closed-form crosstalk noise metrics for physical design applications, in *Proceedings of the Design, Automation and Test Conference* (2002), pp. 812–819
114. H.J. Lee, C.C. Chu, W.S. Feng, Crosstalk estimation in high speed VLSI interconnect using coupled RLC tree models, in *Asia-Pacific Conference Circuits and Systems* (2002), pp. 257–262
115. A. Nieuwoudt, J. Kawa, Y. Massoud, Crosstalk-induced delay, noise, and interconnect planarization implications of fill metal in nanoscale process technology. IEEE Trans. Very Large Scale Integr. Syst. **18**, 378–391 (2010)
116. A. Naeemi, R. Venkatesan, J.D. Meindl, Optimal global interconnects for GSI. IEEE Trans. Electron Devices **50**, 980–987 (2003)
117. A. Vittal, L.H. Chen, M.M. Sadowska, K.P. Wang, X. Yang, Modeling crosstalk in resistive VLSI interconnections, in *Proceedings of the 12th International Conference VLSI Design* (1999), pp. 470–475
118. L. Avinash, M.K. Krishna, M.B. Srinivas, A novel encoding scheme for delay and energy minimization in VLSI interconnects with built-in error detection, in *IEEE Computer Society Annual Symposium VLSI* (2008), pp. 128–133
119. E. Nuroska, S.J. Ruun, F. Lai, U. Schwiegelshohn, L.C. Liu, On optimizing power and crosstalk for bus coupling capacitance using genetic algorithms, in *Proceedings of the International Symposium Circuits and Systems* (2003), pp. 277–280
120. N. Hanchate, N. Ranganathan, A linear time algorithm for wire sizing with simultaneous optimization of interconnect delay and crosstalk noise, in *Proceedings of the 19th International Conference VLSI Design* (2006), pp. 283–292
121. J. Lienig, A parallel genetic algorithm for performance-driven VLSI routing. IEEE Trans. Evol. Comput. **1**, 29–39 (1997)
122. R.R. Rao, H.S. Deogun, D. Blaauw, D. Sylvester, Bus encoding for total power reduction using a leakage-aware buffer configuration. IEEE Trans. Very Large Scale Integr. Syst. **13**, 1376–1383 (2005)
123. T. Zhang, S.S. Sapatnekar, Simultaneous shield and buffer insertion for crosstalk noise reduction in global routing. IEEE Trans. Very Large Scale Integr. Syst. **15**, 624–636 (2007)
124. D. Wu, J. Hut, R. Mahapatra, M. Zhao, Layer assignment for crosstalk risk minimization, in *Proceedings of the Asia and South Pacific Design Automation Conference* (2004), pp. 159–162
125. T. Ho, Y. Chang, S. Chen, D.T. Lee, A fast crosstalk-and performance- driven multilevel routing system, in *International Conference Computer Aided Design* (2003), pp. 382–387

126. M. Yoshikawa, H. Terai, Crosstalk-driven placement based on genetic algorithms, in *IEEE International Conference Computational Intelligence for Measurement Systems and Applications* (2004), pp. 70–75
127. K.S. Sainarayanan, J.V. Ravindra, M.B. Srinivas, A novel, coupling driven, low power bus coding technique for minimizing capacitive crosstalk in VLSI interconnects, in *Proceedings of the IEEE International Symposium Circuits and Systems* (2006), pp. 4155–4159
128. R.A. Powers, Batteries for low power electronics, in *Proceedings of the IEEE* (1995), pp. 687–693
129. A.P. Chandrakasan, S. Sheng, R.W. Brodersen, Low-power CMOS digital design. IEEE J. Solid-State Circuits **27**, 473–484 (1992)
130. B. Davari, R.H. Dennard, G.G. Shahidi, CMOS scaling for high performance and low power-the next ten years, in *Proceedings of the IEEE* (1995), pp. 595–606
131. J.D. Meindl, Low power microelectronics: retrospect and prospect, in *Proceedings of the IEEE* (1995), pp. 619–635
132. A.B. Bhattacharyya, R.S. Rana, S.K. Guha, R. Bahl, S. Anand, M.J. Zarabi, P.A. Govindacharyulu, U. Gupta, V. Mohan, J. Roy, A. Atri, A micropower analog hearing aid on low voltage CMOS digital process, in *Proceedings of the 9th International Conference VLSI Design* (1996), pp. 85–89
133. P. Corbishley, E. Rodriguez-Villegas, C. Toumazou, An ultra-low power analogue directionality system for digital hearing aids, in *Proceedings of the International Symposium Circuits and Systems* (2004), pp. 233–236
134. F.N. Najm, A survey of power estimation techniques in VLSI circuits. IEEE Trans. Very Large Scale Integr. Syst. **2**, 446–455 (1994)
135. S.S. Rajput, S.S. Jamuar, High current, low voltage current mirrors and applications, in *Proceedings of the 10th International Conference VLSI* (1999), pp. 47–60
136. S.S. Rajput, S.S. Jamuar, Low voltage analog circuit design techniques. IEEE Circuits Syst. Mag. **2**, 24–42 (2002)
137. R.X. Gu, M.I. Elmasry, Power dissipation analysis and optimization of deep submicron CMOS digital circuits. IEEE J. Solid-State Circuits **31**, 707–713 (1996)
138. M. Borah, R.M. Owens, M.J. Irwin, Transistor sizing for minimizing power consumption of CMOS circuit under delay constraint, in *Proceedings of the International Symposium Low Power Design* (1995), pp. 167–172
139. L.S. Heulser, W. Fichtner, Transistor sizing for large combinational digital CMOS circuits. Integr. VLSI J. **10**, 185–212 (1991)
140. A.P. Chandrakasan, R.W. Brodersen, Minimizing power consumption in digital CMOS circuits, in *Proceedings of the IEEE* (1995), pp. 498–523
141. S.M. Kang, Y. Leblebici, *CMOS Digital Integrated Circuits-Analysis and Design* (McGraw Hill, New York, 2003)
142. A.P. Chandrakasan, R.W. Brodersen, *Sources of Power Consumption in Low Power Digital CMOS Design* (Kluwer, Norwell, 1995)
143. J.M. Rabaey, M. Pedram, *Low Power Design Methodologies* (Kluwer, New York, 2002)
144. S.M. Kang, Accurate simulation of power dissipation in VLSI circuits. IEEE J. Solid-State Circuits **SC-21**, 889–891 (1986)
145. G.Y. Yaccub, W.H. Ku, An enhanced technique for simulating short-circuit power dissipation. IEEE J. Solid-State Circuits **24**, 844–847 (1989)
146. G. Constandinou, J. Georgiou, C. Toumazou, Nano-power mixed-signal tunable edge detection circuit for pixel-level processing in next generation vision systems. Electron. Lett. **39**, 1774–1775 (2003)
147. C. Kim, I.C. Hwang, S. Kang, A low-power small-area ±7.28ps jitter 1-GHz DLL-based clock generator. IEEE J. Solid-State Circuits **37**, 1414–1420 (2002)
148. B. Bhaumik, P. Pradhan, G.S. Visweswaran, R.Varambally, A. Hardi, A low power 256 KB SRAM design, in *Proceedings of the 12th International Conference VLSI Design* (1999), pp. 67–70

149. S. Mitra, A.N. Chandorkar, Design of amplifier with rail-to-rail CMR with 1V power supply, in *Proceedings of the 17th International Conference VLSI Design* (2004), pp. 52–56
150. I.C. Hwang, C. Kim, S.M. Kang, A CMOS self-regulating VCO with low supply sensitivity. IEEE J. Solid-State Circuits **39**, 42–48 (2004)
151. A. Lidow, D. Kinzer, G. Sheridan, D. Tam, The semiconductor roadmap for power management in the new millennium, in *Proceedings of the IEEE* (2001), pp. 803–812
152. A.J. Bhavnagarwala, B.L. Austin, K.A. Bowman, J.D. Meindl, A minimum total power methodology for projecting limits on CMOS GSI. IEEE Trans. Very Large Scale Integr. Syst. **8**, 235–251 (2000)
153. S. Mutoh, T. Douseki, Y. Matsuya, T. Aoki, S. Shigematsu, J. Yamada, 1-V power supply high speed digital circuit technology with multithreshold voltage CMOS. IEEE J. Solid-State Circuits **30**, 847–854 (1995)
154. H. Kawaguchi, K. Nose, T. Sakurai, A CMOS scheme for 0.5V supply voltage with pico-ampere standby current, in *IEEE International Solid State Circuits Conference* (1998), pp. 192–193
155. L. Wei, Z. Chen, K. Roy, M.C. Johnson, Y. Ye, V.K. De, Design and optimization of dual-threshold circuits for low-voltage low-power applications. IEEE Trans. Very Large Scale Integr. Syst. **7**, 16–24 (1999)
156. A.R. Khalid, R. Paily, FPGA implementation of high speed and low power architectures for image segmentation using SOBEL operators. J. Circuits Syst. Comput. **21**, 16–29 (2012)
157. P. Pant, V. De, A. Chatterjee, Simultaneous power supply, threshold voltage, and transistor size optimization for low-power operation of CMOS circuits. IEEE Trans. Very Large Scale Integr. Syst. **6**, 538–545 (1998)
158. J.C. Chi, H.H. Lee, S.H. Tsai, M.C. Chi, Gate level multiple supply voltage assignment algorithm for power optimization under timing constraint. IEEE Trans. Very Large Scale Integr. Syst. **15**, 637–648 (2007)
159. V.V. Deodhar, J.A. Davis, Optimization of throughput performance for low-power VLSI interconnects. IEEE Trans. Very Large Scale Integr. Syst. **13**, 308–318 (2005)
160. R. Chandel, S. Sarkar, R.P. Agarwal, An analysis of interconnect delay minimization by low-voltage repeater insertion. Microelectron. J. **38**, 649–655 (2007)
161. R. Chandel, S. Sarkar, R.P. Agarwal, Transition time considerations in voltage-scaled repeaters. Microelectron. Int. **22**, 39–40 (2005)
162. K. Banerjee, A. Mehrotra, A power-optimal repeater insertion methodology for global interconnects in nanometer designs. IEEE Trans. Electron Devices **49**, 2001–2007 (2002)
163. P. Wang, G. Pei, E.C. Kan, Pulsed wave interconnect. IEEE Trans. Very Large Scale Integr. Syst. **12**, 453–463 (2004)
164. L. Zhong, N.K. Jha, Interconnect-aware low-power high-level synthesis. IEEE Trans. Comput. Aided Des. Integr. Circuits Syst. **24**, 336–351 (2005)
165. A. Tajalli, Y. Leblebici, Design trade-offs in ultra-low power digital nanoscale CMOS. IEEE Trans. Circuits Syst. I Regul. Pap. **58**, 2189–2200 (2011)
166. R.H. Reuss, M. Fritze, Introduction to special issue on circuit technology for ULP, in *Proceedings of the IEEE* (2010), pp. 139–143
167. S.D. Pable, M. Hasan, High speed interconnect through device optimization for subthreshold FPGA. Microelectron. J. **42**, 545–552 (2011)
168. D. Bol, R. Ambroise, D. Flandre, J.D. Legat, Interests and limitations of technology scaling for subthreshold logic. IEEE Trans. Very Large Scale Integr. Syst. **17**, 1508–1519 (2009)
169. B.H. Calhoun, D.C. Daly, N. Verma, D.F. Finchelstein, D.D. Wentzloff, A. Wang, S. Cho, A.P. Chandrakasan, Design considerations for ultra-low energy wireless microsensor nodes. IEEE Trans. Comput. **54**, 727–740 (2005)
170. H. Soeleman, Ultra-low power digital sub-threshold logic design, Ph.D. Dissertation, Purdue University, USA, 2000
171. G. Schrom, S. Selberherr, Ultra-low-power CMOS technologies, in *International Semiconductor Conference* (1996), pp. 237–246

172. B.H. Calhoun, A. Chandrakasan, Modeling and sizing for minimum energy operation in subthreshold circuits. IEEE J. Solid-State Circuits **40**, 1778–1786 (2005)
173. R.M. Swanson, J.D. Meindl, Ion-implanted complementary MOS transistors in low-voltage circuits. IEEE J. Solid-State Circuits **SC-7**, 146–153 (1972)
174. E. Vittoz, J. Fellrath, CMOS analog integrated circuits based on weak-inversion operation. IEEE J. Solid-State Circuits **12**, 224–231 (1977)
175. C. Mead. *Analog VLSI and Neural Systems* (Addison-Wesley, Reading, 1989)
176. R. Lyon. C. Mead, An analog electronic cochlea. IEEE Trans. Acoust. Speech Signal Process. **36**, 1119–1134 (1988)
177. S. Hanson, B. Zhai, S. Mingoo, B. Cline, K. Zhou, M. Singhal, M. Minuth, J. Olson, L. Nazhandali, T. Austin, D. Sylvester, D. Blaauw, Exploring variability and performance in a sub-200 mV processor. IEEE J. Solid-State Circuits **43**, 881–891 (2008)
178. B. Zhai, S. Hanson, D. Blaauw, D. Sylvester, A variation-tolerant sub-200 mV 6-T subthreshold SRAM. IEEE J. Solid-State Circuits **43**, 2338–2348 (2008)
179. J. Kwong, Y.K. Ramadass, N. Verma, A.P. Chandrakasan, A 65 nm sub-Vt microcontroller with integrated SRAM and switched capacitor DC-DC converter. IEEE J. Solid-State Circuits **44**, 115–126 (2009)
180. B.S. Chaurasia, S. Tandon, S. Shukla, P. Mishra, A. Mohan, S.K. Balasubramanium, Modelling and simulation of blast wave for pressure sensor design. Int. J. Adv. Eng. Technol. **1**, 64–71 (2011)
181. G.K. Prasad, J.S. Sahambi, Classification of ECG arrhythmias using multimedia-resolution analysis and neural networks, in *Proceedings of the IEEE Conference Convergent Technologies* (2003), pp. 227–231
182. B.C. Paul, H. Soeleman, K. Roy, An 8 × 8 sub-threshold digital CMOS carry save array multiplier, in *Proceedings of the 27th European Solid-State Circuits Conference* (2001), pp. 377–380
183. B. Zhai, L. Nazhandali, J. Olson, A. Reeves, M. Minuth, R. Helfand, S. Pant, D. Blaauw, T. Austin, A 2.60pJ/inst subthreshold sensor processor for optimal energy efficiency, in *Symposium VLSI Circuits, Digest of Technical Papers* (2006), pp. 154–155
184. A. Wang, A. Chandrakasan, A 180-mV subthreshold FFT processor using a minimum energy design methodology. IEEE J. Solid-State Circuits **40**, 310–319 (2005)
185. H. Soeleman, K. Roy, B.C. Paul, Robust subthreshold logic for ultra-low power operation. IEEE Trans. Very Large Scale Integr. Syst. **9**, 90–99 (2001)
186. H. Soeleman, K. Roy, B.C. Paul, Sub-domino logic: ultra-low power dynamic sub-threshold digital logic, in *14th International Conference VLSI Design* (2001), pp. 3–7
187. M. Anis, M.H. Aburahma, Leakage current variability in nanometer technologies, in *Proceedings of the 5th International Workshop System-on-Chip for Real-Time Applications* (2005), pp. 60–63
188. T. Xinghai, V.K. De, J.D. Meindl, Intrinsic MOSFET parameter fluctuations due to random dopant placement. IEEE Trans. Very Large Scale Integr. Syst. **5**, 369–376 (1997)
189. W. Shockley, Problems related to pn junctions in silicon. Solid-State Electron. **2**, 35–60 (1961)
190. R.W. Keyes, The effect of randomness in the distribution of impurity atoms on FET thresholds. IEEE J. Solid-State Circuits **10**, 245–247 (1975)
191. J. Kwong, A. Chandrakasan, Variation-driven device sizing for minimum energy subthreshold circuits, in *Proceedings of the International Symposium Low Power Electronics and Design* (2006), pp. 8–13
192. L.P. Melek, M.C. Schneider, C. Galup-Montoro, Body-bias compensation technique for subthreshold CMOS static logic gates, in *17th Symposium Integrated Circuits and Systems Design* (2004), pp. 267–272
193. B. Zhai, S. Hanson, D. Blaauw, D. Sylvester, Analysis and mitigation of variability in subthreshold design, in *Proceedings of the International Symposium Low Power Electronics and Design* (2005), pp. 20–25

194. T. Kim, H. Eom, J. Keane, C. Kim, Utilizing reverse short channel effect for optimal subthreshold circuit design, in *Proceedings of the International Symposium Low Power Electronics and Design* (2006), pp. 127–130
195. R. Ramirez, J. Jaffari, M. Anis, Variability aware design of subthreshold devices, in *IEEE International Symposium Circuits and Systems* (2008), pp. 1196–1199
196. A. Srivastava, D. Sylvester, D. Blaauw, *Statistical Analysis and Optimization for VLSI: Timing and Power* (Springer, New York, 2005)
197. R. Rao, A. Srivastava, D. Blaauw, D. Sylvester, Statistical analysis of leakage current for VLSI circuits. IEEE Trans. Very Large Scale Integr. Syst. **12**, 131–139 (2004)
198. K. Agarwal, S. Nassif, Characterizing process variation in nanometer CMOS, in *Proceedings of the 44th ACM/IEEE Design Automation Conference* (2007), pp. 396–399
199. P. Stolk, F.P. Widdershoven, D.M. Klaassen, Modeling statistical dopant fluctuations in MOS transistors. IEEE Trans. Electron Devices **45**, 1960–1971 (1998)
200. S. Borkar, T. Karnik, S. Narendra, J. Tschanz, A. Keshavarzi, V. De, Parameter variations and impact on circuits and microarchitecture, in *Proceedings of the Design Automation Conference* (2003), pp. 338–342
201. T.A. Brunner, Impact of lens aberrations on optical lithography. IBM J. Res. Dev. **41**, 57–67 (1997)
202. A.K. Wong, R.A. Ferguson, S.M. Mansfield, The mask error factor in optical lithography. IEEE Trans. Semicond. Manuf. **13**, 235–242 (2000)
203. A.R. Alvarez, L.A. Akers, Monte Carlo analysis of sensitivity of threshold voltage in small geometry MOSFETs. Electron. Lett. **18**, 42–43 (1982)
204. L.-O. Bauer, M.R. MacPherson, A.T. Robinson, H.G. Dill, Properties of silicon implanted with boron ions through thermal silicon dioxide. Solid-State Electron. **16**, 289–300 (1973)
205. W. Schemmert, G. Zimmer, Threshold-voltage sensitivity of ion-implanted MOS transistors due to process variations. Electron. Lett. **10**, 151–152 (1974)
206. K. Kuhn et al., Managing process variation in Intels 45 nm CMOS technology. Intel Technol. J. **12**, 93–109 (2008)
207. N. Verma, A. Chandrakasan, A 65 nm 8T sub-Vt SRAM employing sense-amplifier redundancy, in *IEEE International Solid-State Circuits Conference* (2007), pp. 328–329
208. B. Datta, W. Burleson, Temperature effects on energy optimization in sub-threshold circuit design, in *10th International Symposium Quality Electronic Design* (2009), pp. 680–685
209. C. Rossi, P. Aguirre, Ultra low-power CMOS cells for temperature sensors, in *Proceedings of the 18th Symposium Integrated Circuits and Systems Design* (2005), pp. 202–206
210. Y.S. Lin, D. Blaauw, D. Sylvester, An ultra low power 1V, 220 nW temperature sensor for passive wireless applications, in *IEEE Custom Integrated Circuits Conference* (2008), pp. 507–510
211. S. Hanson, B. Zhai, M. Seok, B. Cline, K. Zhou, M. Singhal, M. Minuth, J. Olson, L. Nazhandali, T. Austin, D. Sylvester, D. Blaauw, Performance and variability optimization strategies in a sub 200 mV, 3.5 pJ/inst, 11 nW subthreshold processor, in *IEEE Symposium VLSI Circuits* (2007), pp. 152–153
212. K. Rais, F. Bulestra, G. Ghibaudo, Temperature dependence of gate induced drain leakage current in silicon CMOS devices. Electron. Lett. **30**, 32–34 (1994)
213. W. Fikry, G. Ghibaudo, M. Dutoit, Temperature dependence of drain induced barrier lowering in deep sub-micrometer MOSFETs. Electron. Lett. **30**, 911–912 (1994)
214. O. Semenov, A. Vassighi, M. Sachdev, Leakage current in sub-quarter micron MOSFET: a perspective of stressed delta IDDQ testing. J. Electron. Test. Theory Appl. **19**, 341–352 (2003)
215. G. Ghibaudo, F. Balestra, Low temperature characterization of silicon CMOS devices, in *Proceedings of the 20th International Conference Microelectronics* (1995), pp. 613–622
216. J.H. Huang, G.B. Zhang, Z.H. Liu, J. Duster, S.J. Wann, K. Ping, H. Chenming, Temperature dependence of MOSFET substrate current. IEEE Electron Device Lett. **14**, 268–271 (1993)

217. W.D. Liu, Study of NMOSFET substrate current mechanisms in the temperature range of 77-295K, in *4th International Conference Solid-State and Integrated Circuit Technology* (1995), pp. 425–427
218. J.H. Anderson, F.N. Najm, Low power programmable FPGA routing circuitry. IEEE Trans. Very Large Scale Integr. Syst. **17**, 1048–1060 (2009)
219. M. Alioto, Understanding DC behavior of subthreshold CMOS logic through closed-form analysis. IEEE Trans. Circuits Syst. part I **57**, 1597–1607 (2010)
220. V. Adler, E.G. Friedman, Delay and power expressions for a CMOS inverter driving a resistive-capacitive load. Analog Integr. Circ. Sig. Process. **14**, 29–39 (1997)
221. K.T. Tang, E.G. Friedman, Delay and power expressions characterizing a CMOS inverter driving an RLC load, in *Proceedings of the IEEE International Symposium Circuits and Systems* (2000), pp. 4.269–4.272
222. Predictive Technology Model (PTM) (2012), http://ptm.asu.edu
223. H. Li, W.Y. Win, J.F. Mao, Modelling of carbon nanotube interconnects and comparative analysis with Cu interconnects, in *Proceedings of the Asia-Pacific Microwave Conference* (2006), pp. 1361–1364
224. M. Kavicharan, N.S. Murthy, N.B. Rao, Modal decomposition based VLSI interconnect delay modeling, in *International Confernce Solid-State Integrated Circuit* (2012), pp. 23–27
225. F. Dartu, L.T. Pileggi, Calculating worst-case gate delay due to dominant capacitive coupling, in *Proceedings of the IEEE/ACM International Conference Computer-Aided Design* (1997), pp. 46–51
226. W.J. Bowhill et al., Circuit implementation of a 300-MHz 64-bit second-generation CMOS alpha CPU. Digital Tech. J. **7**, 100–118 (1995)
227. T. Xiao, M. Marek-Sadowska, Gate sizing to eliminate crosstalk induced timing violation, in *Proceedings of the IEEE International Conference Computer Design* (2001), pp. 186–191
228. I. Jiang, Y. Chang, J. Jou, Crosstalk-driven interconnect optimization by simultaneous gate and wire sizing. IEEE Trans. Comput. Aided Des. Integr. Circuits Syst. **19**, 999–1010 (2000)
229. K. Bowman, S. Duvall, J. Meindl, Impact of die-to-die and within-die parameter fluctuations on the maximum clock frequency distribution for gigascale integration. IEEE J. Solid State Circuits **37**, 183–190 (2002)
230. Z.H. Liu, C. Hu, J.H. Huang, T.Y. Chan, M.C. Jeng, P.K. Ko, Y.C. Cheng, Threshold voltage model for deep-submicrometer MOSFETs. IEEE Trans. Electron Devices **40**, 86–95 (1993)
231. P. Chen, D.A. Kirkpatrick, K. Keutzer, Miller factor for gate-level coupling delay calculation, in *Proceedings of the IEEE/ACM International Conference Computer-Aided Design* (2000), pp. 68–74
232. J.A. Ayers, *Digital Integrated Circuits-Analysis and Design* (CRC Press, New York, 2004)
233. A.A. Giunta, S.F. Wojtkiewicz Jr., M.S. Eldred, Overview of modern design of experiments methods for computational simulations, in *AIAA Aerospace Sciences Meeting and Exhibit* (2003), pp. 1–7
234. M. Graziano, M.R. Casu, G. Masera, G. Piccinini, M. Zamboni, Effects of temperature in deep-submicron global interconnect optimization in future technology nodes. Microelectron. J. **35**, 849–857 (2004)
235. S.C. Lin, N. Srivastava, K. Banerjee, A thermally-aware methodology for design-specific optimization of supply and threshold voltages in nanometer scale ICs, in *Proceedings of the IEEE International Conference Computer Design* (2005), pp. 411–416
236. T.S. Shelar, G.S. Visweswaran, Inclusion of thermal effects in the simulation of bipolar circuits using circuit level behavioral modeling, in *Proceedings of the 17th International Conference VLSI Design* (2004), pp. 821–826
237. Y.P. Tsividis, *Operating and Modeling of the MOS Transistor* (McGraw-Hill, New York, 1999)
238. M.M. Hossain, C. Shakher, Temperature measurement in laminar free convective flow using digital holography. Appl. Opt. **48**, 1869–1877 (2009)

Zeitfracht Medien GmbH
Ferdinand-Jühlke-Straße 7
99095 Erfurt, Deutschland
produktsicherheit@kolibri360.de